TEACHER GUIDE | Includes Works...

9th–12th Grade | Scie...

Quizzes & Tests

Survey of Science History & Concepts

First printing: March 2017
Second printing: June 2018

For information write:

Master Books®, P.O. Box 726, Green Forest, AR 72638

Master Books® is a division of the New Leaf Publishing Group, Inc.

ISBN: 978-1-68344-065-9
ISBN: 978-1-61458-593-0 (digital)

Unless otherwise noted, Scripture quotations are from the New King James Version of the Bible.

Printed in the United States of America

Please visit our website for other great titles:
www.masterbooks.com

For information regarding author interviews,
please contact the publicity department at (870) 438-5288.

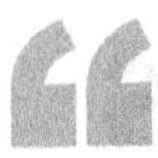

Your reputation as a publisher is stellar. It is a blessing knowing anything I purchase from you is going to be worth every penny!

—Cheri ★★★★★

Last year we found Master Books and it has made a HUGE difference.

—Melanie ★★★★★

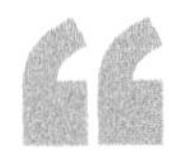

We love Master Books and the way it's set up for easy planning!

—Melissa ★★★★★

You have done a great job. MASTER BOOKS ROCKS!

—Stephanie ★★★★★

Physically high-quality, Biblically faithful, and well-written.

—Danika ★★★★★

Best books ever. Their illustrations are captivating and content amazing!

—Kathy ★★★★★

Affordable
Flexible
Faith Building

Table of Contents

Author Bio:
John Hudson Tiner (*Exploring the World of Mathematics, Exploring the World of Physics*) received five National Science Foundation teaching fellowships during his 12 years as a teacher of mathematics and science that allowed him to study graduate chemistry, astronomy, and mathematics. He also worked as a mathematician and cartographer for the Defense Mapping Agency, Aerospace Center in St. Louis, Missouri.

Tiner has received numerous honors for his writing, including the Missouri Writer's Guild award for best juvenile book for *Exploring the World of Chemistry*. He and his wife, Jeanene, live in Missouri.

Features: The suggested weekly schedule enclosed has easy-to-manage lessons that guide the reading, worksheets, and all assessments. The pages of this guide are perforated and three-hole punched so materials are easy to tear out, hand out, grade, and store. Teachers are encouraged to adjust the schedule and materials needed in order to best work within their unique educational program.

Lesson Scheduling: Students are instructed to read the pages in their book and then complete the corresponding section provided by the teacher. Assessments that may include worksheets, activities, quizzes, and tests are given at regular intervals with space to record each grade. Space is provided on the weekly schedule for assignment dates, and flexibility in scheduling is encouraged. Teachers may adapt the scheduled days per each unique student situation. As the student completes each assignment, this can be marked with an "X" in the box.

	Approximately 30 to 45 minutes per lesson, three to four days a week
	Includes answer keys for worksheets, quizzes, and tests.
	Worksheets for each section
	Quizzes and tests are included to help reinforce learning and provide assessment opportunities.
	Designed for grades 9 to 12 in a one-year course to earn 1 science credit

Course Objectives: Students completing this course will

- Discover how biological classification gives each different type of plant or animal a unique name
- Learn about the differing ways that seeds spread around the world
- Assess what food the body uses for long-term storage of energy
- Explain how biologists learned about the stomach and digestion
- Identify the important invention that caused the world to be divided into time zones
- Evaluate the simple math problem that caused the Mars Climate Orbiter to burn up in the Martian atmosphere
- Determine the common unit of measurement originally based on the distance from the equator to the North Pole
- Compare what Da Vinci's Last Supper and the Parthenon have in common
- Learn about the history of physics from Aristotle to Galileo to Isaac Newton
- Discover how the laws of motion and gravity affect everything from the normal activities of everyday life to launching rockets into space
- Find out why pure gold is not used for jewelry or coins
- Identify Humphry Davy's chemical discoveries and learn how they shortened his life

Course Description

Students will study four areas: mathematics, physics, biology, and chemistry. Students will gain an appreciation for how each subject has affected our lives, and for the people God revealed wisdom to as they sought to understand Creation. Each content area is thoroughly explored, giving students a good foundation in each discipline. It's amazing how ten simple digits can be used in an endless number of ways to benefit man. The development of these ten digits and their many uses is the fascinating story in *Exploring the World of Mathematics*. And in *Exploring the World of Physics*, students find a great tool to have a deeper understanding of the important and interesting ways that physics affects our lives. *Exploring the World of Biology* is a fascinating look at life — from the smallest proteins and spores, to the complex life systems of humans and animals. *Exploring the World of Chemistry* brings science to life and is a wonderful learning tool with many illustrations and biographical information.

Grading Options for This Course

It is always the prerogative of an educator to assess student grades however he or she might deem best. The following is only a suggested guideline based on the material presented through this course:

To calculate the percentage of the worksheets, quizzes, and tests, the educator may use the following guide. Divide total number of questions correct (example: 43) by the total number of questions possible (example: 46) to calculate the percentage out of 100 possible. 43/46 = 93 percent correct.

The suggested grade values are noted as follows: 90 to 100 percent = A; 80 to 89 percent = B; 70 to 79 percent = C; 60 to 69 percent = D; and 0 to 59 percent = F.

First Semester Suggested Daily Schedule

Date	Day	Assignment	Due Date	✓	Grade
		First Semester-First Quarter — *Exploring the World of Mathematics*			
Week 1	Day 1	Review Special Project options on page 14 and plan accordingly.			
	Day 2	Read Pages 4-12 • *Exploring the World of Mathematics* • (EWM)			
	Day 3	Counting Years **Ch1: Worksheet 1** • Page 17 • *Teacher Guide Lesson Plan* • (LP)			
	Day 4	Read Pages 14-22 • (EWM)			
	Day 5				
Week 2	Day 6	Counting the Hours **Ch2: Worksheet 1** • Page 19 • (LP)			
	Day 7	Read Pages 24-34 • (EWM)			
	Day 8	Muddled Measuring **Ch3: Worksheet 1** • Page 21 • (LP)			
	Day 9				
	Day 10				
Week 3	Day 11	Read Pages 36-44 • (EWM)			
	Day 12	Measuring by Metric **Ch4: Worksheet 1** • Page 23 • (LP)			
	Day 13	**Chapters 1-4 Quiz 1** • Page 139 • (LP)			
	Day 14	Read Pages 46-52 • (EWM)			
	Day 15				
Week 4	Day 16	Practical Mathematics **Ch5: Worksheet 1** • Page 25 • (LP)			
	Day 17	Read Pages 54-62 • (EWM)			
	Day 18	The Greek Way With Math **Ch6: Worksheet 1** • Page 27 • (LP)			
	Day 19	Read Pages 64-72 • (EWM)			
	Day 20				
Week 5	Day 21	Names for Numbers **Ch7: Worksheet 1** • Page 29 • (LP)			
	Day 22	Read Pages 74-82 • (EWM)			
	Day 23	Number Patterns **Ch8: Worksheet 1** • Page 31 • (LP)			
	Day 24	**Chapters 5-8 Quiz 2** • Page 141 • (LP)			
	Day 25				
Week 6	Day 26	Read Pages 84-94 • (EWM)			
	Day 27	Endless Numbers **Ch9: Worksheet 1** • Page 33 • (LP)			
	Day 28	Read Pages 96-106 • (EWM)			
	Day 29	Math for Scientists **Ch10: Worksheet 1** • Page 35 • (LP)			
	Day 30				

Date	Day	Assignment	Due Date	✓	Grade
Week 7	Day 31	Read Pages 108-118 • (EWM)			
	Day 32	Pure and Applied Math **Ch11: Worksheet 1** • Page 37 • (LP)			
	Day 33	**Chapters 9-11 Quiz 3** • Page 143 • (LP)			
	Day 34	Read Pages 120-130 • (EWM)			
	Day 35				
Week 8	Day 36	Computing Machines **Ch12: Worksheet 1** • Page 39 • (LP)			
	Day 37	Read Pages 132-140 • (EWM)			
	Day 38	Bits and Bytes **Ch13: Worksheet 1** • Page 41 • (LP)			
	Day 39	Read Pages 142-152 • (EWM)			
	Day 40				
Week 9	Day 41	Math on Vacation **Ch14: Worksheet 1** • Page 43 • (LP)			
	Day 42	**Chapter 12-14 Quiz 4** • Page 147 • (LP)			
	Day 43	Chapter 1-14 Study Day			
	Day 44	**Chapter 1-14 Test 1** • Page 149 • (LP)			
	Day 45				
		First Semester-Second Quarter — ***Exploring the World of Physics***			
Week 1	Day 46	Read Pages 4-12 • Exploring the World of Physics (EWP)			
	Day 47	Motion **Ch1: Worksheet 1** • Page 47 • Lesson Plan • (LP)			
	Day 48	Read Pages 14-22 • (EWP)			
	Day 49	Laws of Motion **Ch2: Worksheet 1** • Page 49 • (LP)			
	Day 50				
Week 2	Day 51	Read Pages 24-32 • (EWP)			
	Day 52	Gravity **Ch3: Worksheet 1** • Page 51 • (LP)			
	Day 53	Read Pages 34-40 • (EWP)			
	Day 54	Simple Machines **Ch4: Worksheet 1** • Page 53 • (LP)			
	Day 55				
Week 3	Day 56	**Chapters 1-4 Quiz 1** • Page 153 • (LP)			
	Day 57	Read Pages 42-52 • (EWP)			
	Day 58	Energy **Ch5: Worksheet 1** • Page 55 • (LP)			
	Day 59				
	Day 60				

Date	Day	Assignment	Due Date	✓	Grade
	Day 61	Read Pages 54-64 • (EWP)			
	Day 62	Heat **Ch6: Worksheet 1** • Page 57 • (LP)			
Week 4	Day 63	Read Pages 66-76 • (EWP)			
	Day 64	States of Matter **Ch7: Worksheet 1** • Page 59 • (LP)			
	Day 65				
	Day 66	**Chapters 5-7 Quiz 2** • Page 155 • (LP)			
	Day 67	Read Pages 78-88 • (EWP)			
Week 5	Day 68	Wave Motion **Ch8: Worksheet 1** • Page 61 • (LP)			
	Day 69				
	Day 70				
	Day 71	Read Pages 90-100 • (EWP)			
	Day 72	Light **Ch9: Worksheet 1** • Page 63 • (LP)			
Week 6	Day 73	Read Pages 102-110 • (EWP)			
	Day 74	Electricity **Ch10: Worksheet 1** • Page 65 • (LP)			
	Day 75				
	Day 76	**Chapters 8-10 Quiz 3** • Page 157 • (LP)			
	Day 77	Read Pages 112-122 • (EWP)			
Week 7	Day 78	Magnetism **Ch11: Worksheet 1** • Page 67 • (LP)			
	Day 79				
	Day 80				
	Day 81	Read Pages 124-134 • (EWP)			
	Day 82	Electromagnetism **Ch12: Worksheet 1** • Page 69 • (LP)			
Week 8	Day 83	Read Pages 136-142 • (EWP)			
	Day 84				
	Day 85				
	Day 86	Nuclear Energy **Ch13: Worksheet 1** • Page 71 • (LP)			
	Day 87	Read Pages 144-152 • (EWP)			
Week 9	Day 88	Future Physics **Ch14: Worksheet 1** • Page 73 • (LP)			
	Day 89	**Chapters 11-14 Quiz 4** • Page 161 • (LP)			
	Day 90	**Chapters 1-14 Test 1** • Page 163 • (LP)			
		Mid-Term Grade			

Second Semester Suggested Daily Schedule

Date	Day	Assignment	Due Date	✓	Grade
		Second Semester-Third Quarter — ***Exploring the World of Biology***			
Week 1	Day 91	Read Pages 6-14 • *Exploring the World of Biology* (EWB)			
	Day 92	The Hidden Kingdom **Ch1: Worksheet 1** • Page 77 • (LP)			
	Day 93	Read Pages 16-26 • (EWB)			
	Day 94				
	Day 95				
Week 2	Day 96	The Invisible Kingdom **Ch2: Worksheet 1** • Page 79 • (LP)			
	Day 97	Read Pages 28-36 • (EWB)			
	Day 98	Exploring Biological Names **Ch3: Worksheet 1** • Page 81 • (LP)			
	Day 99				
	Day 100	**Chapters 1-3 Quiz 1** • Page 167 • (LP)			
Week 3	Day 101	Read Pages 38-46 • (EWB)			
	Day 102	Growing a Green World **Ch4: Worksheet 1** • Page 83 • (LP)			
	Day 103	Read Pages 48-54 • (EWB)			
	Day 104	Food for Energy and Growth **Ch5: Worksheet 1** • Page 85 • (LP)			
	Day 105				
Week 4	Day 106	Read Pages 56-64 • (EWB)			
	Day 107	Digestion **Ch6: Worksheet 1** • Page 87 • (LP)			
	Day 108	Read Pages 66-72 • (EWB)			
	Day 109	Plant Innovators **Ch7: Worksheet 1** • Page 89 • (LP)			
	Day 110				
Week 5	Day 111	**Chapters 4-7 Quiz 2** • Page 169 • (LP)			
	Day 112	Read Pages 74-82 • (EWB)			
	Day 113	Insects **Ch8: Worksheet 1** • Page 91 • (LP)			
	Day 114				
	Day 115				
Week 6	Day 116	Read Pages 84-90 • (EWB)			
	Day 117	Spiders & Other Arachnids **Ch9: Worksheet 1** • Page 93 • (LP)			
	Day 118	Read Pages 92-98 • (EWB)			
	Day 119	Life in Water **Ch10: Worksheet 1** • Page 95 • (LP)			
	Day 120				

Date	Day	Assignment	Due Date	✓	Grade
Week 7	Day 121	Read Pages 100-108 • (EWB)			
	Day 122	Reptiles **Ch11: Worksheet 1** • Page 97 • (LP)			
	Day 123	**Chapters 8-11 Quiz 3** • Page 171 • (LP)			
	Day 124	Read Pages 110-118 • (EWB)			
	Day 125				
Week 8	Day 126	Birds **Ch12: Worksheet 1** • Page 99 • (LP)			
	Day 127	Read Pages 120-130 • (EWB)			
	Day 128	Mammals **Ch13: Worksheet 1** • Page 101 • (LP)			
	Day 129	Read Pages 132-140 • (EWB)			
	Day 130				
Week 9	Day 131	Frauds, Hoaxes, & Wishful Thinking **Ch14: Worksheet 1** • Page 103 • (LP)			
	Day 132	**Chapters 12-14 Quiz 4** • Page 173 • (LP)			
	Day 133	Study Day Chapter 1-14			
	Day 134				
	Day 135	**Chapters 1-14 Test 1** • Page 175 • (LP)			
		Second Semester-Fourth Quarter — ***Exploring the World of Chemistry***	37 days		
Week 1	Day 136	Read Pages 4-10 • Exploring the World of Chemistry (EWC)			
	Day 137	Ancient Metals **Ch1: Worksheet 1** • Page 107 • (LP)			
	Day 138	Read Pages 12-16 • (EWC)			
	Day 139	The Money Metals **Ch2: Worksheet 1** • Page 109 • (LP)			
	Day 140				
Week 2	Day 141	Read Pages 18-26 • (EWC)			
	Day 142	The Search for Gold **Ch3: Worksheet 1** • Page 111 • (LP)			
	Day 143	Read Pages 28-34 • (EWC)			
	Day 144	Gases in the Air **Ch4: Worksheet 1** • Page 113 • (LP)			
	Day 145				
Week 3	Day 146	**Chapters 1-4 Quiz 1** • Page 179 • (LP)			
	Day 147	Read Pages 36-42 • (EWC)			
	Day 148	Electricity to the Rescue **Ch5: Worksheet 1** • Page 115 • (LP)			
	Day 149	Read Pages 44-50 • (EWC)			
	Day 150				

Date	Day	Assignment	Due Date	✓	Grade
Week 4	Day 151	Search for Order **Ch6: Worksheet 1** • Page 117 • (LP)			
	Day 152	Read Pages 52-60 • (EWC)			
	Day 153	Sunlight Shows the Way **Ch7: Worksheet 1** • Page 119 • (LP)			
	Day 154	Read Pages 62-68 • (EWC)			
	Day 155				
Week 5	Day 156	The Electron Shows the Way **Ch8: Worksheet 1** • Page 121 • (LP)			
	Day 157	**Chapters 5-8 Quiz 2** • Page 181 • (LP)			
	Day 158	Read Pages 70-76 • (EWC)			
	Day 159	Compounds by Electric Attraction **Ch9: Worksheet 1** • Page 123 • (LP)			
	Day 160				
Week 6	Day 161	Chapter 10: Water-Read Pages 78-84			
	Day 162	Water **Ch10: Worksheet 1** • Page 125 • (LP)			
	Day 163	Read Pages 86-94 • (EWC)			
	Day 164	Carbon and Its Compounds **Ch11: Worksheet 1** • Page 127 • (LP)			
	Day 165				
Week 7	Day 166	Read Pages 96-102 • (EWC)			
	Day 167	Organic Chemistry **Ch12: Worksheet 1** • Page 129 • (LP)			
	Day 168	**Chapters 9-12 Quiz 3** • Page 183 • (LP)			
	Day 169	Read Pages 104-110 • (EWC)			
	Day 170				
Week 8	Day 171	Nitrogen and Its Compounds **Ch13: Worksheet 1** • Page 131 • (LP)			
	Day 172	Read Pages 112-118 • (EWC)			
	Day 173	Silicon and Its Compounds **Ch14: Worksheet 1** • Page 133 • (LP)			
	Day 174	Read Pages 120-126 • (EWC)			
	Day 175				
Week 9	Day 176	Modern Metals **Ch15: Worksheet 1** • Page 135 • (LP)			
	Day 177	Read Pages 128-134 • (EWC)			
	Day 178	Chemistry in Today's World **Ch16: Worksheet 1** • Page 137 • (LP)			
	Day 179	**Chapters 13-16 Quiz 4** • Page 185 • (LP)			
	Day 180	**Chapters 1-16 Test 1** • Page 187 • (LP)			
		Final Grade			

Special Projects

The Exploring series offers a unique perspective filled with biographical, historical, and scientific perspectives. By highlighting the work and relevance of scientists and innovators, students are introduced to the people behind the knowledge and discoveries that continue to impact their world. This provides exceptional learning opportunities above and beyond the worksheets, quizzes, and tests. Below are three areas of possible activities or bonus point projects that can be undertaken to enhance study.

Biographical

- Select your favorite scientist mentioned in the book and do a research paper on this person's life and/or work. Be sure to include details that enhance the understanding of why they worked in the area of science that they chose, information on their worldview (Christian or secular) and why their work remains relevant.
- There have been some amazing discoveries by women — see if you can find three discoveries by researching at your local library or online at parent-approved sites.

Historical

- Do three short essays — no more than two typed pages each — on discoveries that laid the groundwork for future science fields or the advancement of knowledge.
- Discover where 25 important discoveries related to mathematics or science took place; mark the map for each place and label with the name of each discovery.
- The Bible contains some amazing mathematical and scientific information. Using the geneaological information in Genesis 5, see if you can calculate how many years took place between creation and the Flood of Noah.

Scientific

- Imagine an invention related to mathematics, biology, chemistry, or physical science that could change the way you and others live. See if you can visualize your invention by drawing it out or providing details that would enable someone else to understand the relevance of your invention and how it works.

Applied Learning

These ideas provide a way for the student to acquire knowledge and then apply it — whether that is done in a technical sense or by being able to recognize the concepts at work in the course of their daily experiences. Consider doing one of the two following options as an opportunity to earn bonus points or to extend the learning process:

- Take a spiral notebook and name it "My Learning Observations." Then, using the following concepts, mark the date and time you observe each example over a two-week period. Remember, science is happening around you all the time in every day life, so make sure your observations correlate with mathematics, biology, chemistry, or physical science.
- You can keep a running study journal using the words and people to know during your study. By writing down the definition of words, or the contribution of an individual, you can develop a deeper understanding of the subject matter and have notes available when studying for quizzes and exams.

Mathematics Worksheets

for Use with

Exploring the World of Mathematics

Answer T or F for true or false, fill in the blank, or select the letter for the phrase that best completes the sentence.

T F 1. The extra day, or leap day, every four years was put in the calendar to honor Augustus Caesar.

T F 2. The Gregorian calendar has 100 leap days every 400 years.

3. What is the main reason to have leap days?

A B C D 4. The first calendar with a leap day every four years was the one
A. authorized by Julius Caesar
B. used by the American colonies after 1752
C. used by the Babylonians
D. used by the Egyptians

Matching

5.______ day — a. due to the tilt of the earth's axis, equal to three months

6.______ week — b. earth revolves around the sun once

7.______ month — c. earth rotates on its axis once

8.______ season — d. moon revolves around the earth once

9.______ year — e. seven days

Try Your Math

10. The Bible says that Methuselah died at age 969 years (Gen. 5:27). What would be that age in days? (Ignore leap years.)

11. Using the Babylonian calendar of 360 days in a year, how many days are in one-third of a year; one-fifth of a year; one-twentieth of a year; one-sixtieth of a year?

12. Find the population of your city and calculate how many people are likely to have a birthday on February 29.

Answer T or F for true or false, or select the letter for the phrase that best completes the sentence.

A B C D 1. The shortest naturally occurring period of time that ancient people could observe was the (A. day, B. hour, C. week, D. year).

T F 2. The Egyptians divided daylight into 8 or 12 hours depending on whether it was winter or summer.

A B C D 3. The inventors of the hourglass were the (A. Babylonians, B. British Navy, C. Egyptians, D. Romans).

A B 4. A watch with a sweep second hand is known as (A. an analog, B. a digital) watch.

T F 5. Meridians are imaginary lines going around the earth parallel to the equator.

A B C D 6. Military time has hours numbered from 0000 to (A. 0400, B. 1200, C. 2400, D. 3600).

A B C D 7. Time zones were introduced when it became common to travel by (A. airplanes, B. ox carts, C. ships, D. trains).

A B 8. The International date line is in the (A. Atlantic, B. Pacific) ocean.

A B 9. Atomic clocks proved that the earth's rotation (A. is, B. is not) uniform.

A B C 10. The United States became an independent nation in 1776. In 1976, the country celebrated the fact that the United States was two (A. decades, B. centuries, C. millenniums) old.

Try Your Math

11. Assume that the first four-hour watch began at midnight. What time would it be at five bells on the second watch?

12. Feel your pulse at the wrist and count the number of beats in a minute. Calculate the number of times your heart beats in a day.

13. An office job is often described as working from 9 to 5. This means 9:00 a.m. to 5:00 p.m. How many hours is this?

14. At 4:00 p.m., a family on vacation drives from Mountain Standard Time into Central Standard Time. Should their watches be set one hour earlier to 3:00 p.m. or one hour later to 5:00 p.m.?

Answer T or F for true or false, fill in the blank, or select the letter for the phrase that best completes the sentence.

A B C D 1. NASA's Climate Orbiter to Mars failed because (A. American and French engineers did not communicate with one another, B. engineers used two different measures of force, C. fuel had been measured improperly, D. the spacecraft weighed too much).

A B 2. A troy ounce was used to measure (A. small and expensive, B. large and inexpensive) items.

A B C D 3. A scruple was a standard of weight for measuring (A. barley, B. diamonds, C. drugs, D. potatoes).

A B C D 4. At first, the United States Customary system agreed with that of (A. Britain, B. France, C. Morocco, D. Spain).

T F 5. The American ton and the British tonne are identical in weight.

A B C D 6. Most early measures of distance were based on (A. animal strides, B. human body, C. parts of ships, D. Roman military terms).

7. The length of a mile in feet is ______________.

8. "A pint is a ______________ the world around."

Choose the larger:

A B 9. A. foot, B. yard

A B 10. A. fathom, B. yard

A B 11. A. nautical mile, B. statute mile

A B 12. A. cup, B. quart

A B 13. A. bushel, B. peck

Try Your Math

14. Recall that a hand is four inches. How tall is a horse in inches that is 15 hands tall? How tall in feet?

15. Change your weight from pounds to ounces.

16. The tallest mountain on earth is Mt. Everest. Its summit is 29,035 feet above sea level. How high is the mountain in miles?

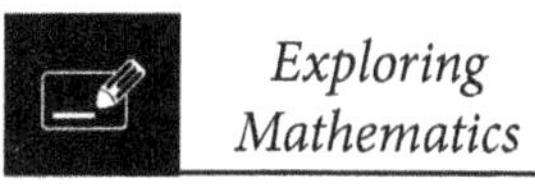	*Exploring Mathematics*	Measuring by Metric p. 36–44	Day 12	Chapter 4 Worksheet 1	Name

Answer T or F for true or false, or select the letter for the phrase that best completes the sentence.

A B 1. The metric system began in (A. Britain, B. France).

A B C 2. The metric system is based on powers of (A. 2, B. 10, C. 12).

T F 3. The metric system was designed specifically to meet the needs of merchants.

A B C D 4. Currently, the meter is defined as (A. 1,640,763.73 wavelengths of krypton gas, B. $\frac{1}{10,000,000}$ of the distance from the equator to the North Pole, C. the distance between two scratch marks on a metal rod, D. the distance light travels in $\frac{1}{299,792,458}$ of a second).

T F 5. Volume (capacity) is a derived unit because it is based on a container that is $\frac{1}{10}$ of a meter on each side.

A B C 6. One meter is slightly longer than one (A. inch, B. yard, C. mile).

A B C 7. One liter is slightly larger than one (A. pint, B. quart, C. gallon).

A B C 8. One kilogram is about 2.2 times as much as (A. one ounce, B. one pound, C. one ton).

A B 9. A standard kilogram is defined by (A. the mass of a platinum cylinder, B. the wavelength of krypton gas).

A B C D 10. Daniel Fahrenheit set the boiling temperature of water on his thermometer at (A. 0, B. 32, C. 100, D. 212) degrees.

A B C D 11. Most people liked Fahrenheit thermometers because (A. they were free, B. they were accurate, C. Fahrenheit was an Englishman, D. daytime temperatures stayed between 0 and 100 degrees).

T F 12. The metric system is illegal to use in the United States.

A B C 13. The (A. Celsius, B. Fahrenheit, C. Kelvin) temperature scale starts at absolute zero.

Answer T or F for true or false, or select the letter for the phrase that best completes the sentence.

A B C D 1. The country known as "the gift of the Nile" was (A. Burundi, B. Egypt, C. Rwanda, D. Sudan).

T F 2. One of the reasons the Egyptians developed mathematics was to figure taxes.

T F 3. Geometry means "to measure a pyramid."

A B 4. The long side of a right triangle is known as the (A. leg, B. hypotenuse).

A B C D 5. The Egyptian knotted rope was used to measure out (A. a pyramid with sloping sides, B. a rectangle with parallel sides, C. a silo of a fixed height, D. a triangle with a right angle).

T F 6. A quadrilateral and triangle are two names for the same figure.

A B C 7. Square feet is an example of a measure of (A. area, B. distance, C. volume).

A B C D 8. Doubling the length, width, and height of a box gives it a volume (A. twice as great, B. three times as great, C. six times as great, D. eight times as great).

A B C D 9. The distance around a circle is called its (A. circumference, B. diameter, C. height, D. radius).

A B 10. The expressions 22 to 7, 22/7, 3.14, and π all refer to the ratio of the circumference of a circle to its (A. diameter, B. radius).

Matching

11._____ circle
12._____ pentagon
13._____ rectangle
14._____ right triangle
15._____ square

a. a polygon with five sides
b. a rectangle with four equal sides
c. a polygon with three sides and one right angle
d. a quadrilateral with opposite sides parallel and equal in length
e. is not a polygon

Matching Formulas

16._____ 2L + 2W
17._____ 4S
18._____ Ah
19._____ L x W
20._____ πr^2
21._____ S^2

a. area of a circle
b. area of a rectangle
c. area of a square
d. perimeter of a rectangle
e. perimeter of a square
f. volume

22. A room is 10 feet wide and 14 feet long. How many square tiles, one foot on a side, would be needed to completely cover the room?

Answer T or F for true or false, or select the letter for the phrase that best completes the sentence.

A B 1. The (A. Egyptians, B. Greeks) strove to understand the principles of mathematics.

2. The sum of the ______________of the legs of a right triangle are equal to the ______________ of the hypotenuse.

A B C 3. The figure that encloses the greatest area with the least perimeter or circumference is the (A. circle, B. square, C. triangle).

A B C D 4. A whispering gallery has a shape like (A. a circle, B. a hyperbola, C. a parabola, D. an ellipse).

A B 5. If an object follows an elliptical orbit, then it is on (A. a closed, B. an open) path.

Matching

6.______ Archimedes

7.______ Euclid

8.______ Johannes Kepler

9.______ Pythagoras

10.______ Thales

a. discovered that the sum of the three angles of any triangle is 180 degrees

b. used ratios to find the heights of buildings

c. proved planets follow elliptical orbits

d. wrote *Elements of Geometry*

e. ancient Greek who worked out a way to show large numbers that he called myriads

Matching

11.______ circle	a. a mirror of this shape will focus sunlight
12.______ ellipse	b. all points are the same distance from the center
13.______ parabola	c. the first part of the name means over or beyond
14.______ hyperbola	d. the orbit of Halley's comet is of this shape

Answer T or F for true or false.

T F 1. Some cultures counted with 20 as the base.

T F 2. Any mark that is used to stand for a number is called a digit.

T F 3. The value of a number does not depend on how it is represented.

T F 4. Place value gives a symbol a different value depending upon its location.

T F 5. The digit 0 was invented at the same time as the digits 1 through 9.

T F 6. The digit 0 first came into use in India.

T F 7. Italian merchants packaged goods by the dozen because the number 12 could be divided into smaller portions.

T F 8. Mathematics is sometimes called the ruler of science.

T F 9. Isaac Newton introduced the use of place value and the numeral 0 to Europe.

T F 10. Fibonacci wrote a book called *Elements of Counting.*

T F 11. The prefix bi means one-half.

T F 12. The word billion has the same meaning in England as in the United States.

T F 13. Of the prefixes giga, mega, and tera, the one that has the greatest value is mega.

Answer T or F for true or false, or select the letter for the phrase that best completes the sentence.

A B C D 1. The sentence, "Madam, I'm Adam" is an example of a (A. composite statement, B. palindrome, C. permutation, D. Roman oration).

A B C D 2. The study of the properties of whole numbers is called (A. algebra, B. geometry, C. number theory, D. real analysis).

A B 3. The general form of an even number is (A. 2n, B. 2n + 1), with *n* a whole number.

A B 4. Two is a factor of all (A. even, B. odd) numbers.

A B 5. The number with the greater number of factors is (A. 12, B. 13).

A B C 6. Nine is an example of (A. a prime, B. an even, C. an odd) number.

A B 7. An example of a composite number is (A. 11, B. 12).

A B C D 8. Prime numbers can be found with the sieve of (A. Eratosthenes, B. Euclid, C. Gauss, D. Pythagoras).

A B 9. As you count higher and higher, prime numbers become (A. more and more common, B. rarer and rarer).

T F 10. The statement "A composite number can be written as the product of prime numbers in only one way" has not yet been proven to be true.

T F 11. The statement "Every even number greater than two is the sum of two primes" has not yet been proven to be true.

A B 12. Encrypted data (A. is especially easy for any computer to read and display, B. can only be read by the sender and receiver).

A B C D 13. Fibonacci numbers could also be called (A. calculating with an abacus, B. the problem of adding Adder snakes, C. the problem of multiplying rabbits, D. the problem of the Leaning Tower of Pisa).

14. The next Fibonacci number after 89 and 144 is ______________.

T F 15. The Fibonacci series of numbers is seldom found in nature.

Matching

16.______ 1881, 121, 1001 — a. Fibonacci numbers

17.______ 2, 3, 5, 7, 11, 13 . . . — b. palindromes

18.______ 1, 1, 2, 3, 5, 8, 13 . . . — c. prime numbers

19.______ 1, 4, 9, 16, 25 . . . — d. square numbers

20.______ 1, 3, 6, 10, 15 . . . — e. triangular numbers

Answer T or F for true or false, or select the letter for the phrase that best completes the sentence.

A B 1. Another name for whole numbers is (A. irrational numbers, B. integers).

A B 2. Mathematicians rely on (A. examples, B. proofs) to show whether a statement is true or false.

T F 3. The sum of two whole numbers is always a whole number.

T F 4. The product of two whole numbers is always a whole number.

T F 5. The quotient of two whole numbers is always a whole number.

T F 6. A plus sign, +, can mean both the operation of addition and that a number is a positive integer.

A B C D 7. Before the invention of calculators, shares were used to reduce the necessity of doing (A. addition, B. subtraction, C. multiplication, D. division).

A B C D 8. The American colonies divided the real, a Spanish coin, into (A. 2, B. 4, C. 8, D. 12) pieces.

A B C D 9. Two bits is equal to (A. 12½, B. 25, C. 50, D. 100) cents.

A B C D 10. A common fraction can be changed into a decimal by dividing the numerator by the (A. denominator, B. greatest common factor, C. least common multiple, D. remainder).

A B 11. The expression ⅔ = 0.666 . . . is an example of a (A. repeating, B. terminating) decimal.

T F 12. Every number can be written as the ratio of two whole numbers.

A B C D 13. The square root of two, is an example of (A. a common fraction, B. an irrational number, C. a rational number, D. a terminating decimal).

A B C D 14. The digits of the square root of two, $\sqrt{2}$, when expressed as a decimal (A. do not repeat, B. do not terminate, C. do not form a pattern, D. all of the above).

Matching

15.______ 1, 2, 3, 4, 5 — a. counting numbers

16.______ 0, 1, 2, 3, 4, 5 — b. even numbers

17.______ 2, 4, 6, 8, 10 — c. integers

18.______ 1, 3, 5, 7, 9 — d. irrational numbers

19.______ -3, -2, -1, 0, +1, +2, +3 — e. odd numbers

20.______ 1/1, ½, ¾, ⅔ — f. rational numbers

21.______ $\sqrt{2}$, π, $(1 + \sqrt{5})/2$ — g. whole numbers

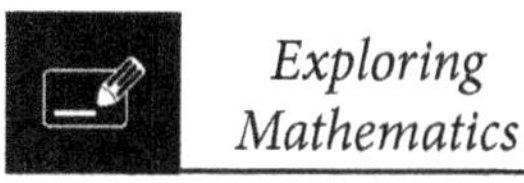

Exploring Mathematics	Math for Scientists p. 96–106	Day 29	Chapter 10 Worksheet 1	Name

Answer T or F for true or false, or select the letter for the phrase that best completes the sentence.

T F 1. Normally, a variable can have more than one value.

A B C D 2. Discovering the value of x when y is equal to zero is called (A. modernizing, B. normalizing, C. solving, D. zeroing) the equation.

A B C D 3. The 5 in the equation $5x + 2 = 0$ is called (A. a coefficient, B. a constant, C. an equalizer, D. a variable).

A B 4. An equation such as $ax^2 + bx + c = 0$ is called an equation of the (A. first, B. second) degree.

T F 5. An equation of the type $y = kx$ is called a linear equation.

T F 6. A scientific law must be stated in the metric system of units to be true.

A B C D 7. The expression $1/x$ is called the (A. base, B. identify, C. reciprocal, D. square root) of x.

A B C D 8. In the 1600s, the French mathematician René Descartes discovered a way to combine algebra with (A. computer programming, B. geometry, C. number theory, D. physics).

A B C 9. The three problems that had resisted solutions since ancient Greek times were trisecting the angle, doubling a cube, and (A. bisecting an angle, B. making a right angle, C. squaring the circle).

Match the equation with the figure on right.

10._____ $y = kx$

11._____ $y = kx^2$

12._____ $y = k\sqrt{x}$

13._____ $y = k/x$

a

b

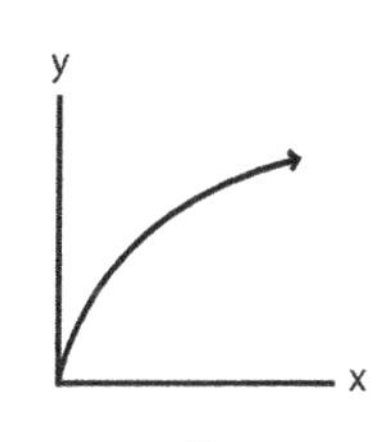

c

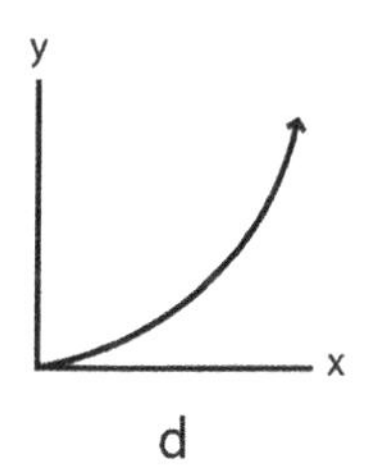

d

Match the statement with the figure above.

14._____ The length a spring stretches (y-axis) is directly proportional to the force pulling on the spring (x-axis).

15._____ The distance an object falls (y-axis) under the influence of gravity is directly proportional to the square of the time it has fallen (x-axis).

16._____ The volume of a gas (y-axis) is inversely proportional to the pressure acting on the gas (x-axis).

17._____ The period of a pendulum (y-axis) is directly proportional to the square root of the length of the pendulum (x-axis).

Answer T or F for true or false, or select the letter for the phrase that best completes the sentence.

A B 1. Mathematics for practical use is called (A. applied, B. pure) mathematics.

A B 2. Discovering large prime numbers to encode data is an example of (A. applied, B. pure) math.

A B C D 3. The problem that a computer helped solve was the (A. bell peal problem, B. binomial theorem, C. four-color map problem, D. Königsberg bridge problem).

A B 4. Each arrangement of the letters ABC, ACB, BAC, BCA, CAB, and CBA is called a (A. combination, B. permutation).

A B C D 5. The expression 3! is read as "three (A. combinations," B. factorial," C. permutations," D. probabilities").

A B C D 6. The value of 5! is (A. 24, B. 25, C. 120, D. 125).

A B 7. To calculate the number of ways that items can be arranged, (A. add, B. multiply) the number of choices for each position.

T F 8. The study of combinations and permutations has no application in everyday life.

Matching

9._____ Andrew Wiles

10._____ Blaise Pascal

11._____ Isaac Newton

12._____ Leonhard Euler

13._____ Pierre de Fermat

a. discovered how to calculate the coefficients of a binomial raised to a power.

b. he called his triangle an arithmetic triangle.

c. his last theorem was solved in 1995.

d. solved the Königsberg bridge problem.

e. proved that $x^n + y^n = z^n$ has no solution with whole numbers except for $n = 2$.

Try Your Math

14. The state of Missouri has license plates with three letters followed by three digits. How many license plates are possible?

Answer T or F for true or false, or select the letter for the phrase that best completes the sentence.

A B C D 1. Any number raised to the zero power is (A. 0, B. 1, C. 2, D. undefined).

T F 2. Babbage completed the analytical engine shortly before his death.

T F 3. Powers of ten can be multiplied by adding their exponents.

T F 4. Fractional exponents are not allowed.

T F 5. A logarithm is an exponent.

6. The expression $\log_{10}3 = 0.477$ is read as "The logarithm of the number __________ in base __________ is __________."

A B C D 7. The number 5,280 changed to standard notation is (A. $.5280 \times 10^1$, B. $5,280 \times 10^3$, C. 5.28×10^2, D. 5.28×10^3).

A B C 8. Scientific measurements are (A. less accurate, B. no more accurate, C. significantly more accurate) than the instruments used to make the measurements.

T F 9. A slide rule multiplies two numbers by adding their logarithms.

A B C D 10. In the early days of computers, input was mainly by (A. colored ribbons, B. punched cards, C. spoken commands, D. switches and relays).

11. In this list, which one is considered the "heart" of a computer: input, control program, memory, central processing unit, output.

A B C D 12. The letters RAM stand for (A. random access memory, B. reasonably accurate member, C. recent abacus modification. D. Robert A. Morley).

Matching

13._____ Augusta Ada Byron, Lady Lovelace	a. built a calculator called the Step Reckoner
14._____ Howard H. Aiken	b. built a calculator to help his father, a tax collector
15._____ Charles Babbage	c. built the first general-purpose calculating machine
16._____ Herman Hollerith	d. built the difference engine
17._____ Johannes Kepler	e. invented logarithms
18._____ Gottfried Leibnitz	f. invented tabulating machines used in the 1890 census
19._____ John Napier	g. spent six years calculating the orbit of Mars
20._____ Blaise Pascal	h. wrote the first computer program

Answer T or F for true or false, or select the letter for the phrase that best completes the sentence.

1. Base 10 uses the digits 0 through ________________.

T F 2. Base 2 uses the digits 0, 1, and 2.

A B C D 3. In computer usage, a single position for a binary digit is called a (A. bit, B. byte, C. kilo, D. pixel).

A B C 4. A single bit can have (A. one, B. two, C. ten) different value(s).

A B C D 5. In personal computers, a byte of data is made of (A. one, B. two, C. eight, D. ten) bit(s).

T F 6. Each character written in ASCII code takes a byte to represent it.

A B C 7. A pixel in a (A. black and white photograph, B. color photograph, C. line drawing) requires the greatest number of bytes.

A B C D 8. In 1861, James Clerk Maxwell made a color photograph using (A. computer enhancement, B. color ink drops sprayed on paper, C. polarized light, D. the three colors of red, green, and blue).

T F 9. Text files cannot be compressed.

T F 10. A pixel is always equal to one bit.

A B C D 11. Video images can be compressed by (A. converting black and white images to color images, B. having reporters avoid standing in front of a blue sky, C. transmitting all pixels that are the same as the previous one, D. transmitting only those pixels that are different from the previous one).

A B C D 12. Moore's law states that computers double in power every 18 (A. days, B. months, C. decades, D. years).

A B C D 13. The bug that Grace Hopper found in the Mark II computer turned out to be (A. a hardware problem, B. a moth caught between mechanical relays, C. a software problem, D. a problem caused by human error).

A B C D 14. A computer with components put farther apart will run more slowly because (A. electric signals can go no faster than the speed of light, B. larger components must be made of less costly materials, C. of resistance in the wires, D. the electrons get lost).

A B C D 15. Engineering students at MIT in the 1950s answered simple questions in computer science with (A. mechanical calculators, B. model cars, C. model trains, D. radio-controlled airplane).

T F 16. The binary digit 1 stands for true, yes, on, or is possible.

A B 17. The query that is most likely to result in more citations is (A. an OR query, B. an AND query).

Try Your Math

18. The Constitution of the United States has 4,609 words and 26,747 characters. At the rate of 7,000 bytes per second, how long would it take a computer to download the Constitution of the United States as an uncompressed text file?

PUZZLES

Puzzle 1: What is so special about 142,857?

The number 142,857 gives interesting results when multiplied by 1 through 6:

1 x 142,857 = 142,857
2 x 142,857 = 285,714
3 x 142,857 = 428,571
4 x 142,857 = 571,428
5 x 142,857 = 714,285
6 x 142,857 = 857,142

The digits in the answer cycle through the digits 142,857 in the same order. You might predict that the answer to 7 x 142,857 would have the same digits. Such a prediction would be wrong. Multiply 7 x 142,857 and see the surprising answer.

Puzzle 2: Multiplying by 99

Some numbers are fun to play around with.

2 x 99 = 198
3 x 99 = 297
4 x 99 = 396
5 x 99 = 495
6 x 99 = 594
7 x 99 = 693
8 x 99 = 792
9 x 99 = 891

The left-most digit (the one in the 100s place) in the answer goes from 1 to 8 while the right-most digit (the one in the 1s place) goes from 8 to 1. Try to figure out the reason for the pattern.

Puzzle 3: On the Road to St. Ives

Try to solve this people on the road puzzle that was turned into an English children's rhyme:

As I was going to St. Ives
I met a man with seven wives;
Every wife had seven sacks;
Every sack had seven cats;
Every cat had seven kits [kittens];
Kits, cats, sacks, and wives,
How many were going to St. Ives?

Can you figure out the answer to the riddle?

Puzzle 4: Send More Money

Here is an addition problem with letters taking the place of numbers. Solve the problem by replacing each letter with one of the digits 0 through 9. Use the same digit for the same letter throughout. In a puzzle like this, it is understood that 0 is not allowed as the first letter in any of the words.

```
   S E N D
 + M O R E
 M O N E Y
```

Hint: Start on the left side. M in MONEY must be 1 because even with a carry, the sum of S and M is less than 20.

Puzzle 5: The 3N + 1 Problem

Pick a number, divide by 2 if it is even. But if it is odd, then multiply by 3 and add 1. Keep on doing this to see where it leads.

For instance, start with 6 (even)

Divide: 6 ÷ 2 = 3 (odd)

Multiply by 3 and add 1: 3 x 3 + 1 = 10 (even)

Divide: 10 ÷ 2 = 5 (odd)

Multiply by 3 and add 1: 3 x 5 + 1 = 16 (even)

Divide: 16 ÷ 2 = 8 (even)

Divide: 8 ÷ 2 = 4 (even)

Divide: 4 ÷ 2 = 2 (even)

Divide: 2 ÷ 2 = 1 (stop)

In every number that mathematicians have tried, the result is always 1. However, no one has yet been able to supply a proof or find an example that does not end with one. Try it with 18.

Puzzle 6: Samson's Riddle

The Bible has puzzles such as Samson's riddle in Judges 14:14: He replied, "Out of the eater, something to eat; out of the strong, something sweet." Hint: You can find the answer in Judges 14:8.

Puzzle 7: Sock Puzzle

Because of an electrical power failure, a boy must get dressed in a dark bedroom. His sock drawer has 10 blue socks and 10 black socks, but in the darkness he cannot tell them apart. He dresses anyway. He reaches into the drawer to grab spare socks so he can change into matching colors later. How many should he take to be certain he has a matching pair?

Puzzle 8: River Crossing

A canoeist must cross a river with three things, but his canoe can hold only one thing at a time. How can the canoeist get a wolf, goat, and carrots across a river? If left alone, the wolf would eat the goat, and the goat would eat the carrots.

Puzzle 9: Durer's Number Square

You can try your hand at making a number square by using the digits 1 through 9 in a three by three square. Each of the rows, columns, and diagonals should add to the same number. Eight different squares are possible.

Puzzle 10: Grass to Milk

Here is a problem to work on a calculator. The answer, when held upside down, shows the name of a four-legged animal that can change green grass into white milk. Find the product of the prime numbers 7, 17, and 23. Add to that answer the area in square feet of a field that is 201 feet on a side, the number of seconds in a day, and 186,000 miles per second (the speed of light). In adding these numbers, ignore the units. Now turn the calculator display around. What name do you see?

Physics Worksheets

for Use with

Exploring the World of Physics

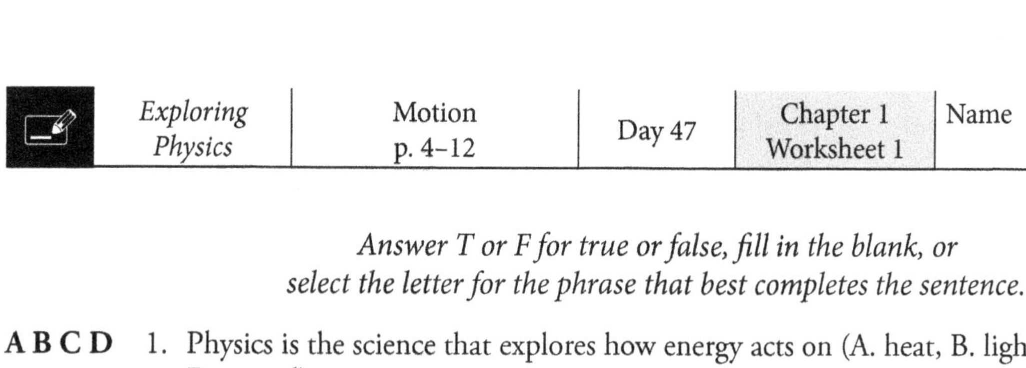

Exploring Physics	Motion p. 4–12	Day 47	Chapter 1 Worksheet 1	Name

Answer T or F for true or false, fill in the blank, or select the letter for the phrase that best completes the sentence.

A B C D 1. Physics is the science that explores how energy acts on (A. heat, B. light, C. matter, D. sound).

T F 2. The ancient Greeks were noted for their careful experiments.

T F 3. The regular back and forth motion of a pendulum was used to regulate the first accurate clocks.

T F 4. In Galileo's time, only length and time could be measured with any accuracy.

A B C D 5. A feather and lump of lead will fall at the same speed in (A. a high speed wind tunnel, B. the atmosphere, C. a vacuum, D. water).

6. To calculate speed, divide distance by ________________.

A B C D 7. To study the motion of falling objects, Galileo (A. beat them into cubes, B. dropped them from a high tower, C. pushed them from a cliff, D. rolled them down a ramp).

A B C D 8. Acceleration is found by dividing the (A. average velocity, B. distance, C. gravity, D. change in speed) by the change in time.

A B 9. On earth, the acceleration due to gravity is (A. 32 ft/sec^2, B. 60 miles/hour).

For More Study

10. Suppose a canoeist takes 70 days to paddle the entire length of the Mississippi River, a distance of 3,710 miles. The canoeist's average speed in miles per day is ________________.

11. An ordinary passenger car can accelerate to 60 miles per hour in about eight seconds. What is the car's acceleration?

12. On the moon, the acceleration due to gravity is 5.3 ft/sec^2 rather than 32 ft/sec^2. If an object fell six seconds before hitting the ground, it strikes the ground with a speed of ________________ ft/sec. (Hint: Use the final velocity equation.)

Answer T or F for true or false, fill in the blank, or select the letter for the phrase that best completes the sentence.

T F 1. Velocity and speed mean the same.

T F 2. A force must act on an object to put the object in motion, give it greater speed, slow it, or change its direction.

T F 3. All objects come to a stop unless some force keeps them going.

T F 4. A ball rolling on a flat surface comes to a stop because of the force of (A. friction, B. gravity).

A B C 5. Isaac Newton's first law of motion was based on experiments done by (A. Aristotle, B. Galileo, C. Newton, himself).

A B C D 6. Inertia is a property of matter that resists changing its (A. electric charge, B. mass, C. momentum, D. velocity).

T F 7. Only very massive objects have inertia.

T F 8. Acceleration is any change of speed or direction.

9. State the second law of motion.

10. State the third law of motion.

11. Momentum is the mass of an object times its ________________.

T F 12. The law of conservation of momentum is one of the most firmly established laws of science.

Matching

13. _____ first law of motion
14. _____ second law of motion
15. _____ third law of motion
16. _____ force equation
17. _____ definition of impulse
18. _____ definition of momentum

a. $a = f/m$
b. $f = m \times a$
c. $f_{ab} = -f_{ba}$
d. $I = f \times t$
e. If $f = 0$ then $a = 0$
f. $p = m \times v$

Answer T or F for true or false, fill in the blank, or select the letter for the phrase that best completes the sentence.

T F 1. During Kepler's time, most people believed the laws governing motions in the heavens differed from those for motions on earth.

A B C D 2. Kepler proved that planets traveled in orbits that were (A. circular, B. elliptical, C. parabolic, D. straight-line).

A B 3. A planet travels (A. faster, B. slower) when closer to the sun.

4. State the second law of planetary motion.

T F 5. Kepler's third law of motion reveals that planets farther from the sun take longer to orbit the sun.

A B 6. Isaac Newton built upon the discoveries of Galileo and (A. Aristotle, B. Kepler).

T F 7. Isaac Newton came from a rich and powerful family.

T F 8. The direction the force of gravity acts on the moon is toward the center of the earth.

A B C D 9. The moon is 60 times as far from the earth as an apple in a tree, so the force of earth's gravity on the moon is (A. 3,600 times weaker, B. 60 times stronger, C. 60 times weaker, D. the same).

T F 10. The law of gravity applies only to the sun, moon, and planets.

A B 11. If the moon were twice as far away, gravitational attraction between the earth and the moon would be (A. one-half B. one-fourth) as great.

12. Force of gravitational attraction between two objects is directly proportional to the ______________ of their masses and inversely proportional to the ______________ of the distance separating them.

T F 13. Scientists have proven that our sun is the only star that has planets orbiting it.

	Exploring Physics	Simple Machines p. 34–40	Day 54	Chapter 4 Worksheet 1	Name

Answer T or F for true or false, fill in the blank, or select the letter for the phrase that best completes the sentence.

1. A simple machine changes the amount of ________________ needed to do a job or the direction the ________________ is applied. (same word)

A B C D 2. The Greek who said, "Give me a place to stand and a long enough lever, and I can move the world" was (A. Archimedes, B. Aristotle, C. Eratosthenes, D. Ptolemy).

A B 3. Mechanical advantage is found by dividing load by (A. effort, B. gravity).

A B C D 4. The tab on a soft drink can is an example of (A. an inclined plane, B. a lever, C. a pulley, D. a wheel and axle).

T F 5. The pivot point (fulcrum) of a lever must be located in the middle.

A B 6. If a load is moved closer to the fulcrum than the effort, the effort required to move the load will be (A. increased, B. reduced).

7. The Grand Canyon is about one mile deep, and the most popular trail out of the canyon is nine miles long; the mechanical advantage of the trail is ________________.

A B C D 8. A screwdriver is an example of (A. a pulley, B. a ramp, C. a wheel and axle, D. an inclined plane).

T F 9. A screw is an inclined plane wrapped around a cylinder.

T F 10. Because of friction, the work produced by a simple machine is greater than the work put into a simple machine.

A B 11. The one that is likely to be the least efficient is (A. a simple machine, B. an 18-wheeler truck).

A B 12. A machine with no friction or other hindrance to its movement would have an efficiency of (A. zero, B. 100) percent.

	Exploring Physics	Energy p. 42–52	Day 58	Chapter 5 Worksheet 1	Name

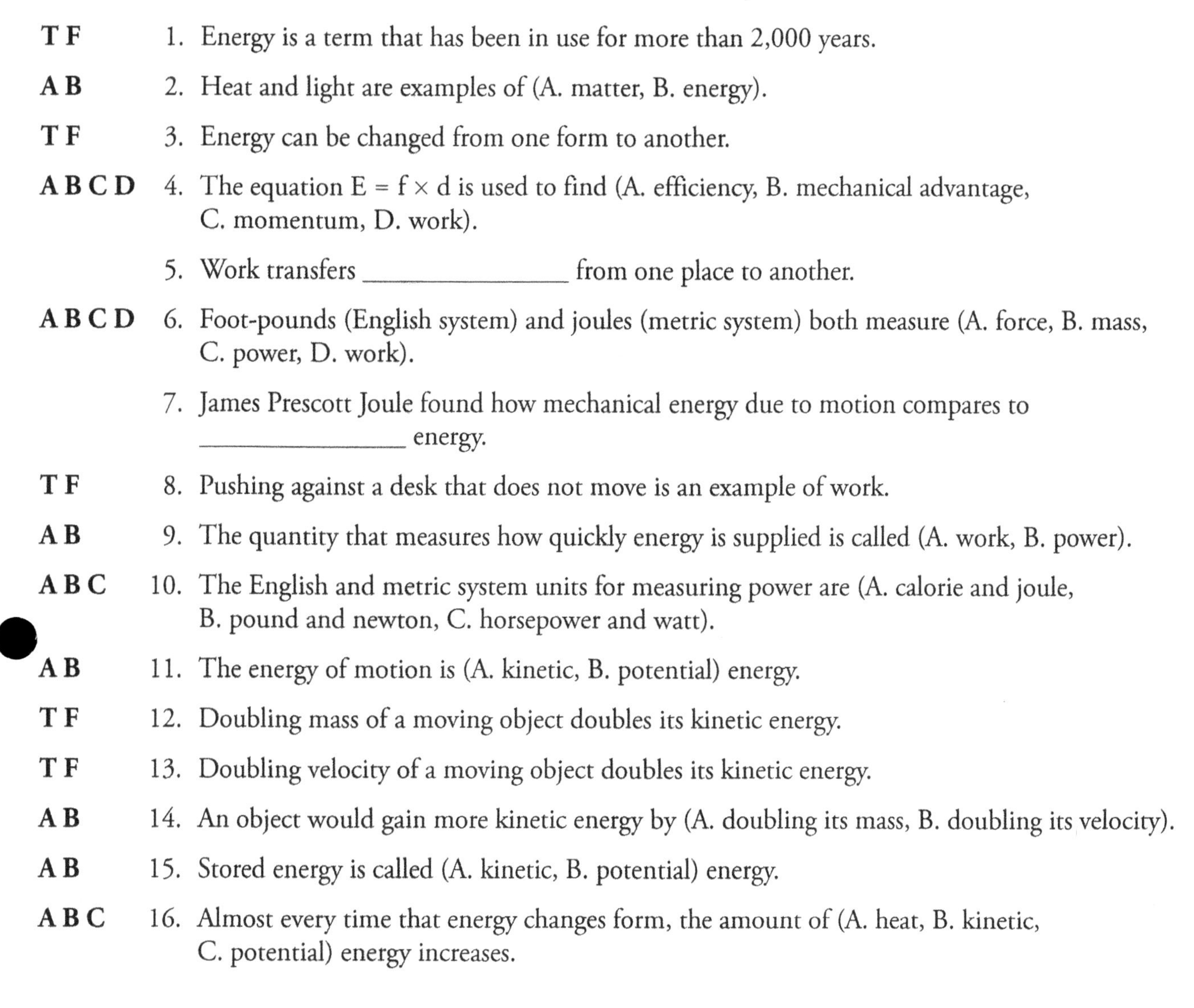

Answer T or F for true or false, fill in the blank, or select the letter for the phrase that best completes the sentence.

T F 1. Energy is a term that has been in use for more than 2,000 years.

A B 2. Heat and light are examples of (A. matter, B. energy).

T F 3. Energy can be changed from one form to another.

A B C D 4. The equation $E = f \times d$ is used to find (A. efficiency, B. mechanical advantage, C. momentum, D. work).

5. Work transfers ________________ from one place to another.

A B C D 6. Foot-pounds (English system) and joules (metric system) both measure (A. force, B. mass, C. power, D. work).

7. James Prescott Joule found how mechanical energy due to motion compares to ________________ energy.

T F 8. Pushing against a desk that does not move is an example of work.

A B 9. The quantity that measures how quickly energy is supplied is called (A. work, B. power).

A B C 10. The English and metric system units for measuring power are (A. calorie and joule, B. pound and newton, C. horsepower and watt).

A B 11. The energy of motion is (A. kinetic, B. potential) energy.

T F 12. Doubling mass of a moving object doubles its kinetic energy.

T F 13. Doubling velocity of a moving object doubles its kinetic energy.

A B 14. An object would gain more kinetic energy by (A. doubling its mass, B. doubling its velocity).

A B 15. Stored energy is called (A. kinetic, B. potential) energy.

A B C 16. Almost every time that energy changes form, the amount of (A. heat, B. kinetic, C. potential) energy increases.

	Exploring Physics	Heat p. 54–64	Day 62	Chapter 6 Worksheet 1	Name

Answer T or F for true or false, fill in the blank, or select the letter for the phrase that best completes the sentence.

A B C D 1. Heat is a type of (A. energy, B. force, C. matter, D. temperature).

2. The three factors that determine the heat contained in an object are type of substance, mass, and ________________.

A B 3. The one that stores heat better is (A. iron, B. water).

A B 4. A thermometer works on the principle that most substances (A. contract, B. expand) when heated.

A B C D 5. The two most common substances used in thermometers are colored alcohol and (A. cooking oil, B. ethylene glycol, C. mercury, D. molten salt).

T F 6. Scientists are unable to measure temperatures greater than 1,700°F.

A B C D 7. The scientist who discovered that pure water has a fixed boiling and freezing temperature was (A. Anders Celsius, B. Antoine Lavoisier, C. Daniel Fahrenheit, D. John Dalton).

A B 8. High air pressure causes water to boil at a (A. higher, B. lower) temperature.

T F 9. Heat is the motion of atoms and molecules.

A B 10. Heat is a form of (A. kinetic, B. potential) energy.

A B C 11. Heat moving from one end of a metal fireplace poker to the other end is an example of heat transfer by (A. conduction, B. convection, C. radiation).

A B 12. The one that conducts heat better is (A. copper, B. wood).

A B 13. Fur, feathers, and other substances with air pockets conduct heat (A. poorly, B. well).

A B C 14. A sea breeze is set in motion because of (A. conduction, B. convection, C. radiation).

T F 15. Heat is transferred from the sun to earth by radiation.

T F 16. A steam engine works because heat flows from a hot region to a cold region.

A B 17. A heat engine works best when the temperature change from heat source to heat sink is (A. about the same, B. greatly different).

T F 18. Moving heat energy in a direction opposite to its normal flow requires work.

For more study

19. The maximum efficiency possible for a machine that produces energy from the difference of ocean water at 18°C at the surface and 1°C at depth is ________________.

	Exploring Physics	States of Matter p. 66–76	Day 64	Chapter 7 Worksheet 1	Name

Answer T or F for true or false, fill in the blank, or select the letter for the phrase that best completes the sentence.

T F 1. A rubber band is elastic because it will stretch.

T F 2. Steel is highly elastic.

3. The amount a solid object bends is directly proportional to the ______________ acting on it.

A B 4. Spreading the weight of a solid over greater area (A. increases, B. reduces) pressure.

A B 5. The factor most important in producing water pressure is the (A. height, B. volume) of the water tank.

T F 6. The pressure of a liquid acts equally in all directions.

A B C D 7. Density is equal to mass divided by (A. area, B. pressure, C. volume, D. weight).

A B 8. As a hurricane approaches, air pressure will (A. increase, B. decrease).

A B C 9. The rate of diffusion of a gas is inversely proportional to the (A. square, B. square root, C. sum) of its molecular weight.

Matching

10.______ Archimedes' principle of buoyancy

11.______ Boyle's law

12.______ Ideal gas law

13.______ Bernoulli's principle

a. Pressure times volume of any gas divided by the temperature is a constant.

b. The lifting force acting on a solid object immersed in water is equal to the weight of the water shoved aside by the object.

c. The velocity of a fluid and its pressure are inversely related.

d. The volume of a gas is inversely proportional to the pressure.

Exploring Physics	Wave Motion p. 78–88	Day 68	Chapter 8 Worksheet 1	Name

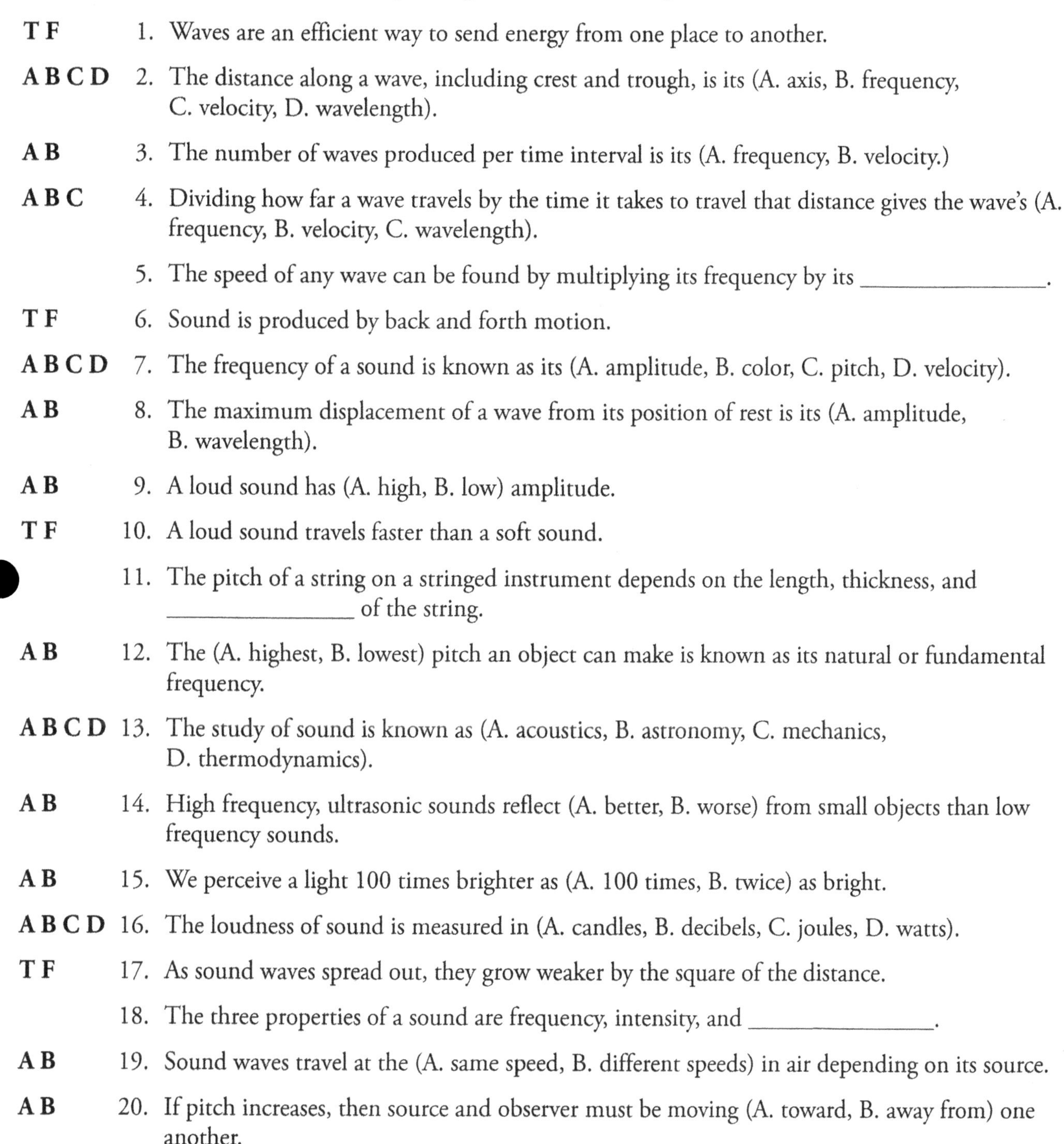

Answer T or F for true or false, fill in the blank, or select the letter for the phrase that best completes the sentence.

T F 1. Waves are an efficient way to send energy from one place to another.

A B C D 2. The distance along a wave, including crest and trough, is its (A. axis, B. frequency, C. velocity, D. wavelength).

A B 3. The number of waves produced per time interval is its (A. frequency, B. velocity.)

A B C 4. Dividing how far a wave travels by the time it takes to travel that distance gives the wave's (A. frequency, B. velocity, C. wavelength).

5. The speed of any wave can be found by multiplying its frequency by its ______________.

T F 6. Sound is produced by back and forth motion.

A B C D 7. The frequency of a sound is known as its (A. amplitude, B. color, C. pitch, D. velocity).

A B 8. The maximum displacement of a wave from its position of rest is its (A. amplitude, B. wavelength).

A B 9. A loud sound has (A. high, B. low) amplitude.

T F 10. A loud sound travels faster than a soft sound.

11. The pitch of a string on a stringed instrument depends on the length, thickness, and ______________ of the string.

A B 12. The (A. highest, B. lowest) pitch an object can make is known as its natural or fundamental frequency.

A B C D 13. The study of sound is known as (A. acoustics, B. astronomy, C. mechanics, D. thermodynamics).

A B 14. High frequency, ultrasonic sounds reflect (A. better, B. worse) from small objects than low frequency sounds.

A B 15. We perceive a light 100 times brighter as (A. 100 times, B. twice) as bright.

A B C D 16. The loudness of sound is measured in (A. candles, B. decibels, C. joules, D. watts).

T F 17. As sound waves spread out, they grow weaker by the square of the distance.

18. The three properties of a sound are frequency, intensity, and ______________.

A B 19. Sound waves travel at the (A. same speed, B. different speeds) in air depending on its source.

A B 20. If pitch increases, then source and observer must be moving (A. toward, B. away from) one another.

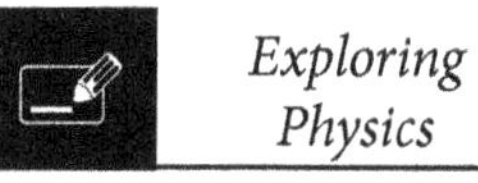

Exploring Physics	Light p. 90–100	Day 72	Chapter 9 Worksheet 1	Name

Answer T or F for true or false, fill in the blank, or select the letter for the phrase that best completes the sentence.

1.______ brings light to a focus. a. Cones

2.______ controls the amount of light that enters the eye. b. Cornea

3.______ is the opening through which light enters the eye. c. Iris

4.______ adjusts light to the best focus. d. Lens

5.______ is a surface of light sensitive nerves. e. Optic nerve

6.______ carries information from the eye to brain. f. Pupil

7.______ is sensitive to light but cannot see color. g. Retina

8.______ is sensitive to light and can distinguish color. h. Rods

T F 9. Sunlight is a mixture of all the colors of the rainbow.

10. The eye has cones that can detect red, green, and ________________ light.

A B 11. The observation that light bounces from a mirror at the same angle at which it enters is known as the law of (A. reflection, B. refraction).

A B 12. The image behind a flat mirror is a (A. real, B. virtual) image.

A B 13. Most modern large telescopes use a (A. lens, B. mirror) to collect light and bring it to a focus.

A B 14. A lens thicker in the middle than at the edges is (A. convex, B. concave).

A B 15. The speed of light is (A. faster, B. slower) in water than in air.

A B 16. The bending of the sun's rays at sunset is an example of (A. refraction, B. reflection).

T F 17. The frequency of light is its brightness.

Exploring Physics	Electricity p. 102–110	Day 74	Chapter 10 Worksheet 1	Name

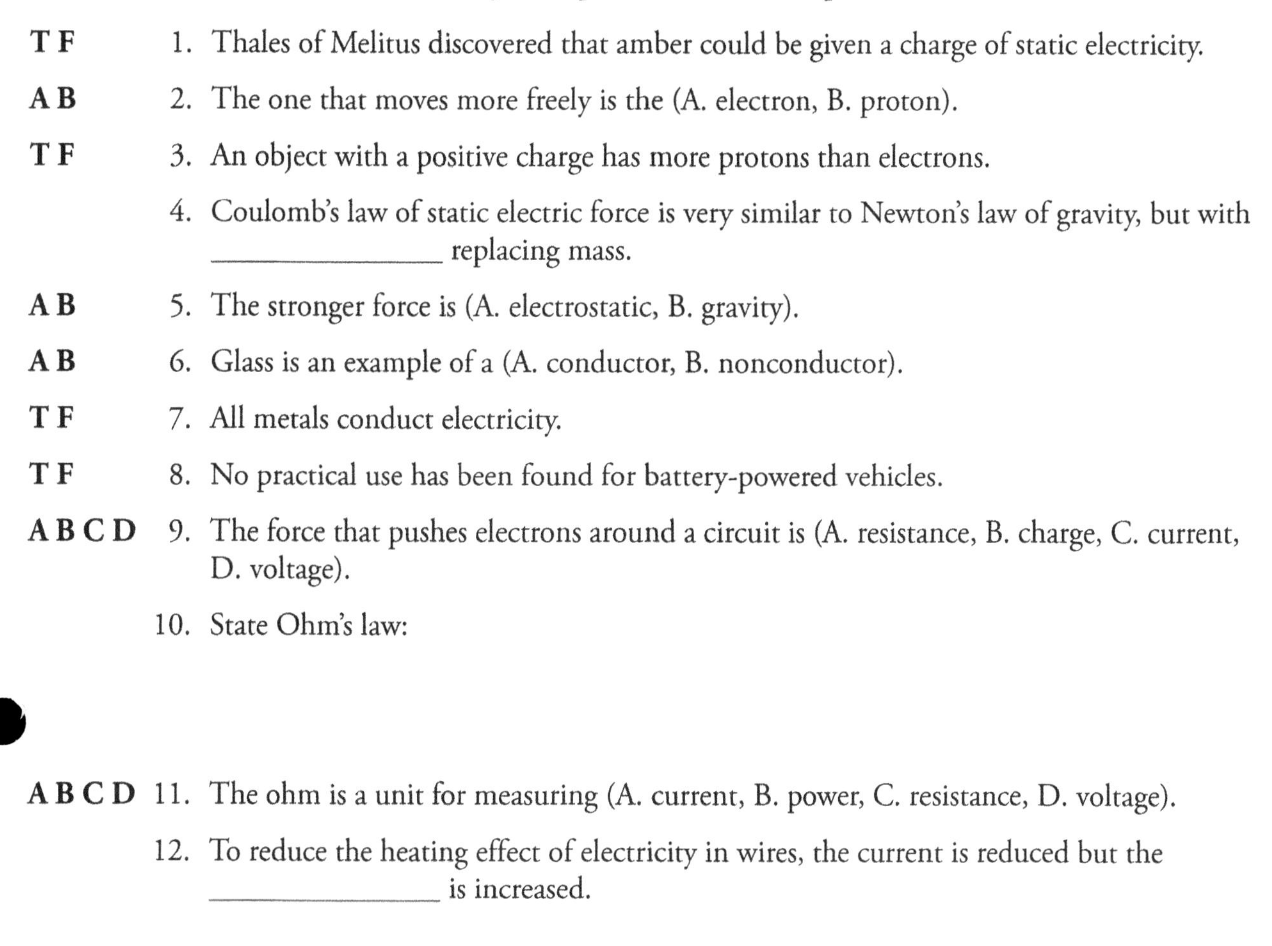

Answer T or F for true or false, fill in the blank, or select the letter for the phrase that best completes the sentence.

T F 1. Thales of Melitus discovered that amber could be given a charge of static electricity.

A B 2. The one that moves more freely is the (A. electron, B. proton).

T F 3. An object with a positive charge has more protons than electrons.

4. Coulomb's law of static electric force is very similar to Newton's law of gravity, but with ________________ replacing mass.

A B 5. The stronger force is (A. electrostatic, B. gravity).

A B 6. Glass is an example of a (A. conductor, B. nonconductor).

T F 7. All metals conduct electricity.

T F 8. No practical use has been found for battery-powered vehicles.

A B C D 9. The force that pushes electrons around a circuit is (A. resistance, B. charge, C. current, D. voltage).

10. State Ohm's law:

A B C D 11. The ohm is a unit for measuring (A. current, B. power, C. resistance, D. voltage).

12. To reduce the heating effect of electricity in wires, the current is reduced but the ________________ is increased.

Exploring Physics	Magnetism p. 112–122	Day 78	Chapter 11 Worksheet 1	Name

Answer T or F for true or false, or select the letter for the phrase that best completes the sentence.

T F 1. Unlike static electricity, magnetism was well understood from the time of the Greeks.

T F 2. William Gilbert proved that the compass is drawn toward the North Star.

T F 3. Earth's geographic North Pole and magnetic north pole have the same location.

A B C 4. The force that can attract but not repel is (A. electric charge, B. gravity, C. magnetism).

T F 5. A negative electric charge can be isolated without a corresponding positive electric charge.

T F 6. Magnets always have a north pole and a south pole.

A B C 7. The metal that a magnet attracts best is (A. aluminum, B. gold, C. iron).

A B 8. Like magnetic poles (A. attract, B. repel) one another.

T F 9. The inverse square law states that a quantity decreases by the square of the distance.

T F 10. Gravity, static electricity, and magnetism all follow the inverse square law.

A B 11. The one that is more difficult to magnetize is (A. soft iron, B. steel).

A B 12. When magnetic domains become jumbled, magnetism is (A. lost, B. strengthened).

T F 13. An electric current can produce a magnetic field.

A B C D 14. The advantage of an electromagnet is that it (A. can be turned on and off, B. does not follow the inverse square law, C. can both attract and repel iron, D. takes less electricity to operate than a natural magnet).

T F 15. Michael Faraday discovered that a moving magnetic field generates electricity.

A B C D 16. Faraday succeeded in showing a connection between (A. chemistry and electricity, B. electricity and magnetism, C. magnetism and light, D. all of the above).

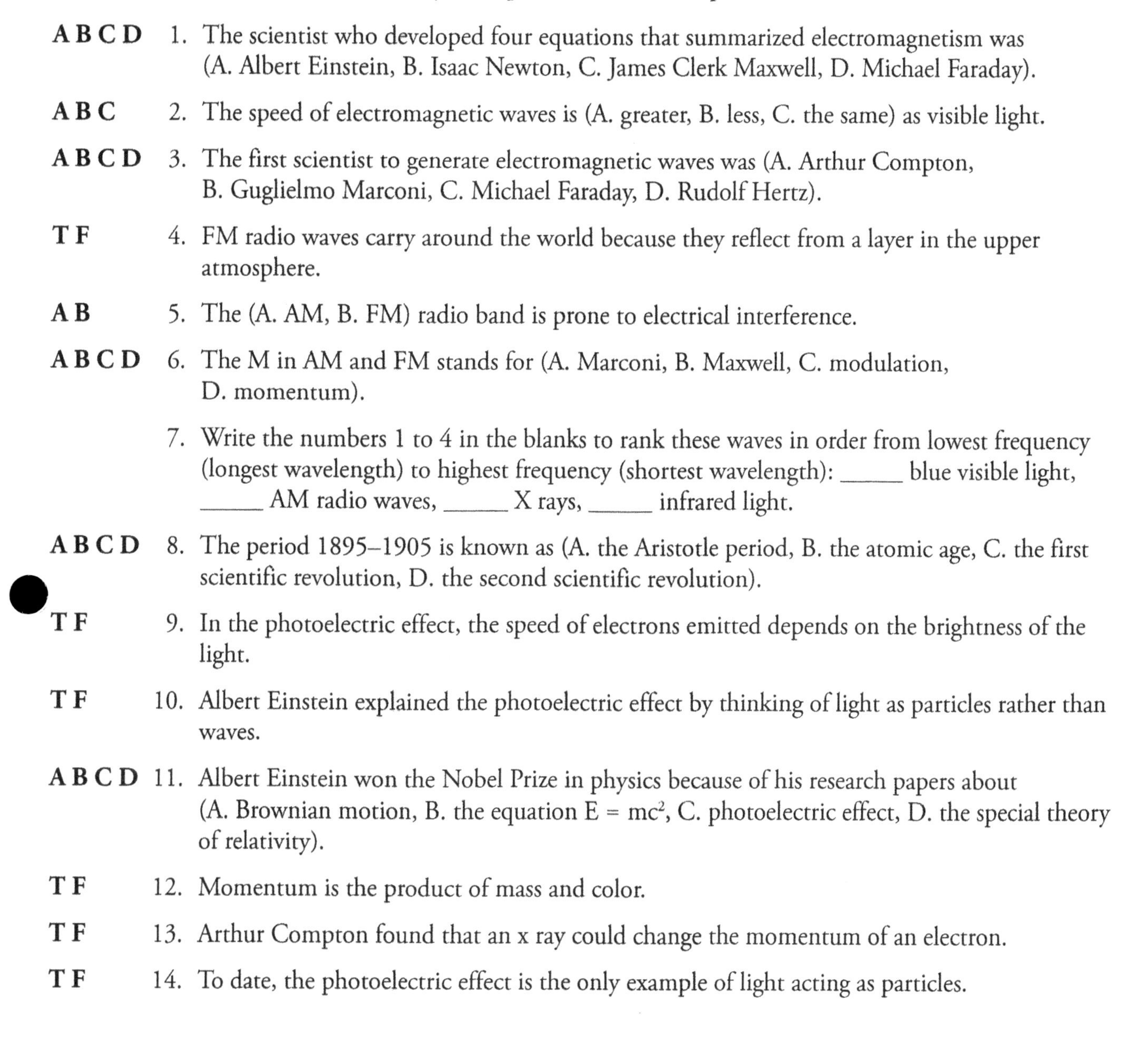

	Exploring Physics	Electromagnetism p. 124–134	Day 82	Chapter 12 Worksheet 1	Name

Answer T or F for true or false, fill in the blank, or select the letter for the phrase that best completes the sentence.

A B C D 1. The scientist who developed four equations that summarized electromagnetism was (A. Albert Einstein, B. Isaac Newton, C. James Clerk Maxwell, D. Michael Faraday).

A B C 2. The speed of electromagnetic waves is (A. greater, B. less, C. the same) as visible light.

A B C D 3. The first scientist to generate electromagnetic waves was (A. Arthur Compton, B. Guglielmo Marconi, C. Michael Faraday, D. Rudolf Hertz).

T F 4. FM radio waves carry around the world because they reflect from a layer in the upper atmosphere.

A B 5. The (A. AM, B. FM) radio band is prone to electrical interference.

A B C D 6. The M in AM and FM stands for (A. Marconi, B. Maxwell, C. modulation, D. momentum).

7. Write the numbers 1 to 4 in the blanks to rank these waves in order from lowest frequency (longest wavelength) to highest frequency (shortest wavelength): _____ blue visible light, _____ AM radio waves, _____ X rays, _____ infrared light.

A B C D 8. The period 1895–1905 is known as (A. the Aristotle period, B. the atomic age, C. the first scientific revolution, D. the second scientific revolution).

T F 9. In the photoelectric effect, the speed of electrons emitted depends on the brightness of the light.

T F 10. Albert Einstein explained the photoelectric effect by thinking of light as particles rather than waves.

A B C D 11. Albert Einstein won the Nobel Prize in physics because of his research papers about (A. Brownian motion, B. the equation $E = mc^2$, C. photoelectric effect, D. the special theory of relativity).

T F 12. Momentum is the product of mass and color.

T F 13. Arthur Compton found that an x ray could change the momentum of an electron.

T F 14. To date, the photoelectric effect is the only example of light acting as particles.

Answer T or F for true or false, or select the letter for the phrase that best completes the sentence.

A B C D 1. The one that is not made of smaller particles is (A. an electron, B. a hydrogen atom, C. a neutron, D. a proton).

A B C 2. The more massive subatomic particle is the (A. electron, B. neutron, C. proton).

A B C 3. A hydrogen atom changes to a helium atom if it gains (A. a neutron, B. a proton, C. an electron).

A B 4. The splitting of an atom into smaller parts is nuclear (A. fission, B. fusion).

T F 5. It is easier to cause the nuclei of atoms to break apart than to melt them together.

A B C D 6. Enrico Fermi discovered that the best particles to cause the break-up of a nucleus were (A. electrons, B. helium nuclei, C. neutrons, D. protons).

A B C D 7. The first person to state that uranium could undergo fission and produce a self-sustaining chain reaction was (A. Albert Einstein, B. Enrico Fermi, C. Franklin D. Roosevelt, D. Lise Meiter).

T F 8. A breeder reactor changes uranium-238 into plutonium.

A B C D 9. The purpose of a moderator is to (A. absorb neutrons, B. cool the reactor, C. generate heat, D. slow neutrons).

A B 10. The total mass after a nuclear reaction is (A. less, B. more) than the total mass before the reaction.

T F 11. More than half of the electricity used in the United States comes from nuclear reactors.

T F 12. The first cold fusion reactor was built in 2002.

	Exploring Physics	Future Physics p. 144–152	Day 88	Chapter 14 Worksheet 1	Name

Answer T or F for true or false, fill in the blank, or select the letter for the phrase that best completes the sentence.

A B 1. In the black box experiment, the amount of ultraviolet light was (A. far less, B. many times greater) than predicted.

T F 2. Max Planck found that the smallest quantum of light was proportional to its speed.

A B 3. An ultraviolet quantum has (A. more, B. less) energy than an infrared quantum.

T F 4. Electrons change orbits only by absorbing or emitting set amounts of energy.

A B 5. The one with a wave long enough to be detected is (A. an electron, B. a baseball).

T F 6. Light waves can interfere with one another.

A B 7. The ground state, or lowest orbit of an electron, corresponds to its (A. fundamental frequency, B. highest overtone).

A B C D 8. The first particle shown to have a wave nature was (A. an alpha particle, B. an electron, C. a gamma ray, D. a proton).

T F 9. The properties that we observe about an electron depend on the experiment that we devise to study the electron.

10. The Heisenberg uncertainty principle states that the precise position, mass, and ________________ for any particle cannot be determined exactly.

Matching

11.______ Niels Bohr

12.______ Louis de Borglie

13.______ Max Planck

14. ______ Werner Heisenberg

a. Developed a model of the atom and electron orbits.

b. Explained black body radiation by using energy quanta.

c. Developed the uncertainty principle.

d. Proposed matter waves and calculated their wavelengths.

Biology Worksheets

for Use with

Exploring the World of Biology

Answer T or F for true or false, or select the letter for the phrase that best completes the sentence.

T F 1. For most of history, living things were classified as either plants or animals.

A B 2. Mushrooms were studied in detail by (A. the Greeks in 400 B.C., B. scientists in the 1700s).

A B C D 3. To keep mushrooms in the plant kingdom, scientists described mushrooms as plants without (A. cell walls, B. chlorophyll, C. seeds, D. sunlight).

A B C D 4. Today, biologists classify mushrooms as members of the (A. animal kingdom, B. bacteria kingdom, C. fungi kingdom, D. plant kingdom).

T F 5. The only way mushrooms can reproduce is by sending out hyphae.

T F 6. The mat of hyphae and the mushrooms it sends to the surface can be one of the largest living things on earth.

A B C D 7. The above-ground stalk and umbrella of a mushroom is used to (A. absorb carbon dioxide, B. catch sunlight, C. release spores, D. sense the presence of enemies).

A B 8. Pigs are used to hunt for (A. truffles, B. death cap mushrooms).

A B 9. Louis Pasteur realized yeast cells were alive when he saw them (A. cause milk to sour, B. grow and reproduce).

A B C D 10. What do yeast cells consume as food? (A. alcohol, B. carbon dioxide, C. sugar, D. vinegar)

A B C D 11. Fungi that are growing on bread and are just visible to the unaided eye and look like miniature mushrooms are most likely: (A. lichen, B. mold, C. truffle, D. yeast).

A B 12. The mold that grew in Alexander Fleming's dish was there because (A. he was experimenting with bread mold, B. it probably drifted in through an open window).

Explore More

Explore More is an opportunity to explore the subject in your own way. Take a photograph, draw a picture, collect a sample, make a poster, write a poem about the subject, list the pros or cons as to whether the subject is helpful or harmful, or interview a person who has experience with the subject. For example, interview a person who has had pneumonia. How did the doctors treat the disease? Have you ever eaten Roquefort cheese? How would you describe its taste?

Subjects for More Exploration

lichen, rust (plant disease), mildew, Dutch elm disease, Roquefort cheese, pneumonia, penicillin-resistant diseases

Answer T or F for true or false, fill in the blank, or select the letter for the phrase that best completes the sentence.

A B 1. The Royal Society employed (A. Robert Hooke, B. Anton van Leeuwenhoek) to test claims of fellow scientists.

A B 2. The first scientist to see the little life in a drop of canal water was (A. Robert Hooke, B. Anton van Leeuwenhoek).

T F 3. The paramecium and amoeba were given the name *protozoa* because they appeared to be animal-like.

A B C D 4. The one that can change its shape is the; (A. amoeba, B. euglena, C. giardiasis, D. paramecium).

T F 5. A protozoa is called simple because it cannot carry out all of life's functions.

6. The single most deadly protozoa disease is ________________.

A B 7. At first, a euglena was called a plant because it (A. could not move, B. had chlorophyll).

A B C D 8. Single-celled algae that surround their cell wall with a coating of silicon dioxide are (A. anaerobic bacteria, B. diatoms, C. giardiasis cysts, D. macrobiotic crust).

T F 9. A member of kingdom Protista must be capable of surviving as a single cell.

10. Lichen is a layer of algae sandwiched between two layers of ________________.

T F 11. All living things must have oxygen to survive.

A B 12. Plants need nitrogen to grow, which they must get from (A. the air, B. nitrogen compounds in the soil.)

Matching

13. a. animal ____ nonliving genetic material that only comes alive inside a living cell

b. bacteria ____ multicellular life that can move and has sense organs

c. fungi ____ multicellular life that includes mushrooms

d. plant ____ multicellular life that makes food by photosynthesis

e. protista ____ single-celled life that includes paramecium, amoeba, and euglena

f. virus ____ single-celled life without a nucleus; one form causes Black Death (plague)

Explore More

Explore More is an opportunity to explore the subject in your own way. View little life through a microscope. Describe what you see. Research the prevention of diseases caused by protozoa, bacteria, and viruses. What are the risks and benefits of protista and bacteria? Read about Robert Hooke and Louis Pasteur. What important discoveries did they make? How does vaccination prevent a disease? What is the difference between the prevention of a disease and the treatment of a disease? How do outdoor experts recommend treating drinking water when backpacking?

	Exploring Biology	Exploring Biological Names, p. 28–36	Day 98	Chapter 3 Worksheet 1	Name

Answer T or F for true or false, fill in the blank, or select the letter for the phrase that best completes the sentence.

1. The Strait of Gibraltar opens from the Mediterranean Sea into the ______________ Ocean.

A B C D 2. The first group of sailors rumored to have discovered the Canary Islands were (A. Greek, B. Phoenician, C. Roman, D. Spanish).

A B C D 3. The Canary Islands were named for (A. birds, B. cats, C. dogs, D. pigs).

T F 4. When a canary began singing loudly, miners knew the air had filled with dangerous gases.

T F 5. *Canis Major* means hot dog.

6. Another name for eyeteeth is ______________ teeth.

A B 7. Aristotle classified dolphins as (A. fish, B. mammals.)

A B C D 8. The English gold coin, the guinea, came from gold found in (A. Guinea along the west coast of Africa, B. Morocco in Northern Africa, C. New Guinea, an island, north of Australia, D. the northern part of South America).

T F 9. Guinea pigs are not pigs but rodents.

T F 10. A guinea pig's front teeth constantly grow.

A B C D 11. The one who wrote *Science of Botany* was (A. a member of the Royal Society, B. Aristotle, C. Carl Linnaeus, D. Plato).

A B 12. The most general name is *kingdom*, but the most specific name is (A. family, B. species).

A B 13. The word *chaos* means (A. disorder or confusion, B. orderly arrangement).

Explore More

The animal kingdom is divided into 33 phyla. Not all of the phyla are well known or contain many species. Explore some of the better-known phyla:

Porifra (sponges)
Coelenterates (jellyfish, hydra, coral, and sea anemones)
Mollusks (clams, oysters, and snails)
Octopuses and squids

Answer T or F for true or false, fill in the blank, or select the letter for the phrase that best completes the sentence.

A B — 1. Most plants produce seeds in (A. the fall, B. early spring).

A B C D — 2. Some plants have seeds with barbs, which are used (A. as a way to be transported elsewhere, B. to protect the growing sprout, C. to provide food for the embryo, D. to delay germination until spring).

3. What nuts were carried along the Silk Road from China to Spain and then to California? ________________________________.

A B C D — 4. What plant has seeds that can be carried thousands of miles by ocean currents? (A. coconut, B. cottonwood, C. dandelion, D. tumbleweed)

T F — 5. The thick leaves of the century plant are storehouses of food energy.

6. The process by which plants can grow from a part of the parent plant such as a cutting from the stem or root is called ________________ reproduction.

A B — 7. Spores are produced by plants that (A. flower, B. do not flower).

T F — 8. A spore is made of a single cell.

9. Plants that take two years to produce seeds are known as ________________.

10. The white potato is what part of the plant? ________________

T F — 11. Photosynthesis produces proteins for growing things.

A B — 12. The drink made from the hard seed of a tree is (A. tea, B. coffee).

Explore More

Visit the produce section of a grocery story. Categorize the produce as vegetables or fruits. Do the vegetables come from the roots, stems, or leaves?

Keep a list of the types of plants that you eat each day. Are they vegetables, fruits, seeds, or grain? How are cornflakes and other breakfast cereals made from the raw crops?

Learn the names of the shrubs, flowers, or trees that grow in a nearby park. Identify each one as annual, biennial, or perennial.

What are other uses for plants besides food? Are plants used to make clothing? How is ethanol made from corn?

Answer T or F for true or false, fill in the blank, or select the letter for the phrase that best completes the sentence.

A B 1. Most of the world's supply of food comes from (A. protein from nuts, B. cereal grains).

T F 2. Corn was brought to the New World by the Spanish explorers.

A B 3. The body can make heat energy by the (A. oxidation, B. reduction) of food.

A B 4. The energy foods are carbohydrates and (A. fats, B. proteins).

A B C D 5. Starch is a type of (A. carbohydrate, B. fat, C. indigestible cellulose, D. protein).

6. The reason cattle can digest grass is because they have ______________ in their digestive system.

A B C D 7. The sugar found in mother's milk is (A. fructose, B. glucose, C. lactose, D. maltose).

8. The simple sugar ready for use by cells is ______________.

9. The three elements found in both carbohydrates and fats are ______________, ______________, and ______________ (any order).

T F 10. The body can convert excess sugar into fat.

A B 11. A calorie is a measure of (A. fat, B. heat energy).

A B C 12. The one that the body uses for long-term storage of energy is: (A. sugar, B. fat, C. protein).

A B C D 13. The one used for growth and repair of the body is (A. carbohydrates, B. fats , C. glycogen, D. proteins).

A B 14. The ones that are made of large molecules are (A. sugars, B. proteins).

Explore More

"Read the label and set a better table" is a ditty that helps people understand what is found in packaged food. In the United States, food packages are labeled with what they contain. Find the food contents labels of several different types of food. Notice the serving sizes. How many servings does each package have? Notice the number of calories per servings. How many grams of fiber and how many grams of fat does each one have? Compare those numbers with several other types of foods. Which ones have the fewest calories, the most fiber, and the least fats?

What is a staple food, and what are the most common staple foods? In what countries is rice the main staple food? Are potatoes a food staple? What was the Potato Famine?

Make a list of foods that contain corn.

What are vitamins, and how does the body use them? What minerals and trace elements does the body require?

	Exploring Biology	Digestion p. 56–64	Day 107	Chapter 6 Worksheet 1	Name

Answer T or F for true or false, fill in the blank, or select the letter for the phrase that best completes the sentence.

A B 1. Teeth are the primary method of (A. chemical, B. mechanical) digestion.

A B 2. Animals with teeth that are broad and flat are probably (A. meat eaters, B. grazing animals).

A B C D 3. Molar teeth are designed to (A. cut food, B. grasp food, C. mash food, D. taste food).

A B 4. Saliva changes carbohydrates into a type of (A. sugar, B. protein).

T F 5. A person can drink, eat, and swallow while standing on his or her head.

A B C D 6. The stomach digests food by (A. electrical impulses similar to a microwave, B. chemically changing the food, C. heating it and softening it, D. mechanically mashing food).

T F 7. All foods are digested at the same rate.

A B 8. The acid found in the stomach to help the action of pepsin is (A. hydrochloric acid, B. sulfuric acid).

A B 9. The one that is longer is the (A. large intestine, B. small intestine).

A B C D 10. The chemical digestion of fats begins in the (A. large intestine, B. mouth, C. small intestine, D. stomach).

A B C D 11. The digested form of protein is (A. amino acids, B. enzymes, C. fatty acids, D. glucose).

A B C D 12. Taste buds are most sensitive to a: (A. bitter taste, B. mustard taste, C. sour taste, D. sweet taste).

Explore More

What is lactose intolerance and how is it treated?

What are essential amino acids?

What is a good source of vitamin A, and what condition does the lack of vitamin A cause? Research the same information for vitamins B_1, C, D, and K.

What minerals are essential in the diet of humans? What condition does a lack of iodine in the diet cause?

Why might some birds find it difficult to swallow food in the weightlessness of a space station?

Answer T or F for true or false, fill in the blank, or select the letter for the phrase that best completes the sentence.

A B 1. While in New England, Luther Burbank took sweet corn to market first because he (A. used a compost hotbed, B. imported them from California).

A B 2. Most farmers plant potatoes by (A. planting potato seeds, B. planting a piece of a potato with an eye in it).

T F 3. Of the 23 potato seeds that Luther Burbank planted, only one had many desirable properties.

4. To what state did Luther Burbank move after leaving New England? ________________

5. Prunes are partially dried fruit from a special type of ________________ tree.

A B 6. To speed up the development of prune trees, Luther Burbank first grew (A. almond trees, B. prune trees) in a greenhouse in a hotbed.

7. John Chapman is better known as Johnny ________________.

8. Why did George Washington Carver do laundry, ironing, and gardening rather than farm work?

A B 9. George Washington Carver became an instructor at: (A. the Naval Academy, B. the Tuskegee Institute in Alabama).

A B C D 10. The four essential elements that plants need are carbon, oxygen, hydrogen, and: (A. iron, B. nitrogen, C. phosphorus, D. sulfur).

11. In Carver's day, the important crops of the South were corn and ________________.

A B C D 12. Carver suggested farmers plant peanuts and sweet potatoes because: (A. nitrogen-fixing bacteria grew along their roots, B. they had fine roots that prevented soil erosion, C. they produced far more income than cotton or corn, D. they released a chemical that killed harmful boll weevils).

T F 13. George de Mestral thought about what caused burrs to hold so strongly while picking them from his cat named Iris.

A B 14. Velcro is also known as a hook and (A. ladder, B. loop) fastener.

Explore More

Who was Louis Agassiz and what are some of his discoveries? Luther Burbank did not study the work of Gregor Mendel on dominant and recessive genes. Research Mendel's discoveries and explain how they could have helped Luther Burbank develop plants with desirable properties. How does one make a compost hotbed? How does a greenhouse provide a warmer temperature for growing plants? What other ways can plants be grown more quickly? What are some of the better-known varieties of potatoes? Are some better for baking than others? How do plant inventors today use artificial selection, cross-pollination, and grafting to produce better fruit and nut trees? Can plants be patented like other inventions? Research the Plant Patent Act and other laws that protect the developer of new plants. What are some of the products that can be made from peanuts and sweet potatoes?

Answer T or F for true or false, fill in the blank, or select the letter for the phrase that best completes the sentence.

Matching

1. Match the number of legs with the type of arthropod.

__ six	a. centipedes
__ eight	b. crabs and lobsters
__ ten	c. insects
__ one pair of legs per body segment	d. millipedes
__ two pair of legs per body segment	e. spiders and ticks

A B 2. The prefix arthro in *arthropod* means (A. foot, B. joint).

A B 3. The one that is a type of insect is (A. cricket, B. shrimp).

T F 4. Biologists often ignore insects that have no obvious role in the daily life of humans.

T F 5. Jean Henri Fabre was the first scientist to bring insects into the laboratory and study them with a microscope.

6. The three divisions of an adult insect's body are ______________, ______________, and ______________.

A B 7. The main goal of adult insects is to (A. eat as much food as possible to survive the winter, B. mate and reproduce).

A B 8. The one with a more noticeable metamorphosis is the (A. butterfly, B. grasshopper).

T F 9. The scientist who discovered the cause of silkworm disease was Jean Henri Fabre.

T F 10. A cicada larva can live underground for as long as 17 years.

A B 11. The one that is a serious plant pest is (A. aphid, B. ladybug).

Explore More

Although Louis Pasteur, the French scientist, began his career as a chemist, he is today remembered for his biological studies. He found the cause of silkworm diseases and how microorganisms can cause food to spoil. He also discovered how to vaccinate against diseases in animals and humans. Explore some of the diseases he studied. What treatment did he recommend for anthrax, chicken cholera, and rabies?

Insects can carry diseases to humans. What insect carries sleeping sickness? Yellow fever? Malaria?

Bees, termites, and ants are called social insects. What is the meaning of that term? How do social insects differ from other insects?

Why do some farmers keep beehives near their crops? Some people say that man's best friend among the insects is the honeybee. What facts support this statement?

Answer T or F for true or false, fill in the blank, or select the letter for the phrase that best completes the sentence.

A B 1. A spider can be described as an arthropod with (A. six legs and three body segments, B. eight legs and two body segments).

T F 2. The most deadly spider to humans is the tarantula.

A B C D 3. The spider that has a distinctive red hourglass pattern on the underside of the abdomen is the (A. female black widow, B. male brown recluse, C. orb weaver, D. tarantula).

A B 4. A brown recluse is most likely to (A. hunt during the day, B. hunt at night).

T F 5. Both spiders and insects are arthropods, but only spiders are arachnids.

A B C D 6. Spider silk is stronger than silkworm silk because spider silk (A. contains a small strand of steel, B. has a different composition, C. is consistent in thickness without weak spots, D. is thicker and shorter).

A B C D 7. The one that is NOT a member of Class Arachnid is (A. grasshopper, B. scorpion, C. spider, D. tick).

8. Why might a cowboy who lives in the desert southwest shake out his boots before putting them on? ______________________________

A B 9. The phrase that describes a parasite is (A. to eat at someone else's table, B. to live in two different places).

A B 10. The one that is considered the most dangerous tick disease today is (A. Lyme disease, B. Rocky Mountain spotted fever).

11. All arthropods have a hard outer covering known as an ____________.

12. Why must arthropods molt? ______________________________

____

Explore More

In addition to the number of legs and number of body segments, insects and spiders differ in the design of their eyes. Explore more about the eyes of insects and of spiders and describe these differences.

Lobsters, crabs, shrimps, crayfish, and barnacles have ten legs (counting two claws) and two body segments. Research these arthropods with a goal of describing their benefit to mankind and in what ways they are detrimental. Identify their habitats, their sources of food, and how their designs help them survive in their environment.

	Exploring Biology	Life in Water p. 92–98	Day 119	Chapter 10 Worksheet 1	Name

Answer T or F for true or false, fill in the blank, or select the letter for the phrase that best completes the sentence.

T F 1. Biologists believe that new plant and animal species are unlikely to be discovered.

A B 2. The one with the greater number of animal species is (A. invertebrates, B. vertebrates).

A B C D 3. The one that is a vertebrate is (A. coral, B. fish, C. lancet, D. snail).

A B C D 4. The classification of vertebrate is a (A. class, B. kingdom, C. phylum, D. subphylum).

A B 5. Fish are (A. cold-blooded, B. warm-blooded).

6. What special sense organ makes it possible for a school of fish to turn together?

____

A B 7. Jacques Cousteau wrote (A. *Twenty Thousand Leagues Under the Sea*, B. *The Silent World*).

A B C D 8. Amphibians include frogs, toads, and (A. catfish, B. goldfish, C. salamanders, D. salmon).

9. Amphibians can breathe through gills, lungs, and _______________.

T F 10. Rather than feathers, wool, or hair, amphibian skin is protected by scales.

A B 11. Amphibians are (A. cold-blooded, B. warm-blooded).

A B 12. Biologists believe the number of amphibians is on the (A. rise, B. decline).

Explore More

What is the deepest part of the ocean? Have explorers managed to dive to that depth? Did they find life there? Choose a hero of undersea exploration and describe his or her discoveries. What are the latest efforts to develop gill backpacks (also called artificial gills) for human underwater swimmers?

Some fish, such as salmon, travel from saltwater to freshwater to spawn. Why do biologists think they make the difficult trip upstream to lay their eggs?

Some people have an aquarium for keeping fish. What must be done to maintain healthy fish in an aquarium? Do you or your friends keep amphibians, reptiles, birds, and mammals as pets?

Fishing is a pleasant hobby for many. What are the most popular game fish? Fishing is also a commercial way of earning a living. What fish are sources of food?

Amphibians include frogs and toads. How do frogs and toads differ from one another?

Answer T or F for true or false, fill in the blank, or select the letter for the phrase that best completes the sentence.

T F 1. Reptiles are vertebrates.

A B 2. Snakes are (A. cold-blooded, B. warm-blooded).

A B C D 3. The one that is not a reptile is a (A. frog, B. lizard, C. snake, D. turtle).

T F 4. Most reptiles have sweat glands to cool themselves.

A B C D 5. The reptile that protects her nest is the (A. crocodile, B. Gila monster, C. sea turtle, D. snake).

6. Why do snakes flick their forked tongues in and out? ______________________________

A B 7. The pit of a copperhead is an organ to sense (A. heat rays, B. odors).

A B C D 8. The recommended treatment for a snake bite is to (A. cut the wound to release venom, B. keep the victim calm and transport him to the hospital, C. use a tight tourniquet to stop the spread of venom, D. use ice to reduce swelling).

A B 9. The one that is more likely to be found in water is the (A. coral snake, B. moccasin).

A B C D 10. The toxin of a coral snake (A. attacks blood vessels, B. blinds a person's vision, c. causes deep puncture wounds, D. damages the nervous system).

A B 11. A python attacks by (A. spitting poison, B. squeezing its victims).

A B 12. Lizards are (A. cold-blooded, B. warm-blooded).

A B C D 13. The poisonous lizard is the (A. chameleon, B. gecko, C. Gila monster, D. Komodo dragon).

14. The lizard that can change color is the _______________.

Matching

15.

___ The most common venomous snake found in the eastern United States	a. anaconda
___ is cottony white inside its mouth	b. copperhead
___ can weigh 20 pounds and is cooked as food	c. coral snake
___ has glossy red, yellow, and black bands along its body	d. diamondback rattlesnake
___ is the largest venomous snake	e. king cobra
___ is a constrictor and is the largest snake by weight	f. moccasin

Answer T or F for true or false, fill in the blank, or select the letter for the phrase that best completes the sentence.

A B 1. Birds are (A. cold-blooded, B. warm-blooded).

A B 2. Compared to other vertebrates, birds have hearts that beat (A. less, B. more)quickly.

A B C D 3. The feature birds do NOT share with other animals is (A. a backbone, B. feathers, C. reproducion by laying eggs, D. walking on two legs).

4. Birds grown for human consumption are referred to as ________________.

5. What birds did the British use to fly messages during the war with Napoleon? ______________________

A B C D 6. The person who made bird watching popular was (A. Carl Linnaeus, B. John J. Audubon, C. Roger Tory Peterson, D. The Duke of Wellington).

A B C D 7. The plumage of a species of a bird can vary depending on whether (A. it is a juvenile, b. it is male or female, C. it is summer or winter, D. all of the above).

A B 8. The brighter and more distinctive plumage usually belongs to the (A. female, B. male) bird.

9. Why do birds eat seeds and insects rather than grasses? ________________________

A B C D 10. An anhinga can sink below the surface easily because (A. it is almost entirely without feathers, B. its feathers are black and heavy, C. its feathers are coated with oil, D. its feathers have no oil coating).

11. Rather than teeth for crushing food, birds have a ________________ filled with grit and small stones.

T F 12. One reason birds migrate is to find food.

Matching

13.

___ Can mimic human speech	a. anhinga
___ Used to hunt small game	b. Arctic tern
___ Thick, heavyset, and flightless, this bird is extinct	c. cassowary
___ Once numerous in the United States, now extinct	d. dodo
___ Has a long, thin beak that is like a hollow straw	e. falcon
___ Has a long, dagger-like beak used to spear fish	f. hummingbird
___ Migrates from Arctic to Antarctic	g. parrot
___ Large bird capable of killing a human	h. passenger pigeon

Explore More

What are some common birds in your area? Can you name them? Watch them to learn about their daily life. Sketch two dissimilar birds and mark how they differ from one another — color, beaks, and type of claws.

The voice box of a bird is called a syrinx. How does it differ from the human voice box? Some birds can achieve a range of tones and have a distinctive call. What birds can you identify from their call or song alone? Some bird-watching books attempt to portray a bird's song as a musical diagram. Study such a diagram for birds whose call you know.

How does the structure of feathers help insulate birds? Why do birds molt?

Explore the flight characteristics of birds. Which ones use thermals to remain aloft? What are some typical speeds at which birds fly when they migrate? Why do geese fly in a V formation? How do they find their way when migrating? How do birds build nests? What do they use? How many eggs do different bird species lay?

Answer T or F for true or false, fill in the blank, or select the letter for the phrase that best completes the sentence.

A B C D 1. The one that is NOT true of mammals is that (A. mammals are cold-blooded, B. mammals are vertebrates, C. mammals have a four-chambered heart, D. mammals have sweat glands).

T F 2. The one feature that is true of all mammals is that the females provide milk for the young.

A B 3. The milk that contains more fat is the milk of (A. horses, B. seals).

T F 4. A four-chambered heart is actually two separate blood pumps.

5. Why are the platypus and spiny anteater, who lay eggs rather than give live birth, classified as mammals? ______________________________

6. The mammal that can fly is the ______________.

A B 7. The word *nocturnal* means (A. active at night, B. hunt by sound echo).

8. An example of a mammal that spends most of its time underground is the ______________________.

T F 9. Whales must come to the surface to breathe air.

A B 10. Animals such as squirrels and beavers are examples of (A. canines, B. rodents).

A B 11. Cats are mammals, in the order carnivore, and belong to the family (A. canine, B. feline).

A B 12. Cheetahs catch their prey primarily by (A. pouncing on them with a sudden leap, B. chasing them down).

A B 13. Elephants are still often referred to as *pachyderms* because of the physical characteristic of (A. thick skin, B. a long, flexible nose).

T F 14. Both sheep and elephants eat grass.

Explore More

All animals except mammals and birds are cold-blooded. What advantage does a warm-blooded design have over a cold-blooded design?

Bears are mammals that hibernate during winter. What is hibernation, and what advantage does it give bears?

Some predators will only eat animals they have killed themselves. But some mammals, such as hyenas, are not as fussy. They will eat carrion, the remains of animals that have died or been killed by other animals. What role do these animals have in nature?

In addition to the platypus and spiny anteater, Australia has a large number of marsupials, such as kangaroos, wombats, koalas, Tasmanian devils, and Tasmanian wolves (extinct). The opossum is the only marsupial that lives in the United States. Investigate these animals to discover how marsupials differ from other mammals.

Answer T or F for true or false, fill in the blank, or select the letter for the phrase that best completes the sentence.

A B 1. A deliberate deception designed to gain money or something of value is a (A. fraud, B. hoax).

A B 2. A paleontologist studies (A. past life, B. how animals migrate).

A B C D 3. The man who glued together parts to make a feathered dinosaur fossil did it (A. to embarrass his employer, B. to gain entry into the United States from China, C. to make the fossil more valuable, D. to prove his skill as a fossil hunter).

A B 4. When Xu Xing, the Chinese scientist, found the other fossil slab in China, he discovered that it (A. did not match, B. was in far better shape) than the one in the United States.

T F 5. The fact that the feathered dinosaur fossil was a fraud was discovered before it was made public.

T F 6. Archaeologists study past human life.

A B C D 7. The Piltdown man deception began with the discovery of (A. broken pottery found in a trunk, B. detailed cave paintings, C. part of a skull, D. the remains of a leg bone).

A B C D 8. The phony bones of Piltdown man were uncovered in (A. England, B. France, C. Germany, D. Spain).

T F 9. Robert Virchow described the bones found in the Neander Valley as being from a short but stout human being.

A B C D 10. The classification of humans, *Homo sapiens*, means: (A. caveman, B. dawn man, C. upright man, D. wise man).

T F 11. The only tools ever found by the Neanderthals were stone clubs.

12. Why did Maria see the painted bulls on the ceiling, but her father had overlooked them?

Explore More — Hoax, Fraud, Fake, or What?

A visitor walked into a small trading post in the desert southwest of the United States. A display showed small pieces of sandstone with colorful designs on them. A sign over the small stones read, "Authentic Native American petroglyphs." A petroglyph (PET-ruh-GLIF) is a carving or drawing on rock. Petroglyphs made by prehistoric Native Americans are often found on government land. Removing them is illegal. Those from private property are rare and often expensive.

If someone had made fake Indian petroglyphs to sell, would the rock drawings be an example of a fraud or hoax?

The petroglyphs on display had a price of only one dollar. The visitor decided to purchase one. When he examined the back of his purchase, he noticed that the artist had signed his name. He looked at others and found autographs on them. In addition, each piece had a year and date written below the name. The date was less than a week old.

The visitor asked the shopkeeper if the small drawings were truly authentic Native American petroglyphs. The shopkeeper said yes, but he smiled and told the visitor to walk outside to the back of the building.

There, in the shade, the visitor saw an art teacher and several Native American children working at a table. They drew the rock paintings.

Question: Do you think the rock drawings qualified as authentic Native American petroglyphs?

Chemistry Worksheets

for Use with

Exploring the World of Chemistry

Answer T or F for true or false, fill in the blank, or select the letter for the phrase that best completes the sentence.

	1. Ancient people hammered the soft pure iron from ______________ into useful tools.
A B C D	2. Charcoal is (A. a meteorite that fell from the heavens, B. a type of coal found in the earth, C. made of almost pure oxygen, D. wood that has been heated without oxygen).
T F	3. The only purpose of carbon in smelting iron from its ore is so it will burn and supply heat.
A B C D	4. Which of these forms of iron is the purest? (A. cast iron, B. charcoal, C. steel, D. wrought iron).
A B	5. Cast iron is (A. brittle and will shatter if struck, B. soft and easily hammered into shape).
A B C D	6. Steel is quenched by (A. burying it in the earth, B. heating it in an oven for several days, C. heating it white hot and thrusting it into cold water, D. raising it overhead for lightning to strike).
	7. Cast iron, steel, and wrought iron differ only in the amount of ______________ they contain.
A B	8. Rusting is a (A. slow, B. rapid) oxidation.
A B	9. A tin can is made mostly of (A. tin, B. steel).
A B C D	10. The one that looks more like silver is (A. brass, B. bronze, C. gold, D. pewter).
T F	11. Metals maintain their properties regardless of temperature.
A B	12. The more expensive metal is (A. aluminum, B. tin).

	Exploring Chemistry	The Money Metals p. 12–16	Day 139	Chapter 2 Worksheet 1	Name

Answer T or F for true or false, fill in the blank, or select the letter for the phrase that best completes the sentence.

1. Gold, silver, and ________________ are known as the coinage metals.

A B C D 2. The first metal mentioned in both the Old and New Testaments is (A. copper, B. gold, C. iron, D. tin).

T F 3. A 14-carat gold ring is pure gold.

T F 4. Gold resists being beaten into thin layers.

T F 5. Pure silver, unlike gold, is hard enough to resist daily wear.

A B C D 6. Bronze and brass are both alloys that contain (A. copper, B. gold, C. iron, D. silver).

A B C D 7. Ancient people made musical instruments of (A. copper alloy, B. iron and mercury, C. sulfur and carbon, D. tin and lead).

A B C D 8. The Statue of Liberty has a skin of (A. copper, B. gold, C. steel, D. zinc).

9. The seven ancient metals are gold, silver, copper, iron, tin, lead, and ____________.

A B C D 10. Another name for mercury is (A. calliston, B. cuprum, C. plumbum, D. quicksilver).

T F 11. A block of lead would float in a pool of mercury.

A B C D 12. The metal used in thermometers and barometers is (A. barium, B. lithium, C. mercury, D. silver).

13. The seven ancient planets (wanderers) are sun, moon, Venus, Jupiter, Mars, Saturn, and ________________.

A B C D 14. Ancient people matched the metal gold with (A. Mars, B. the moon, C. Saturn, D. the sun).

A B C D 15. The Apostle Paul was compared to (A. Mercury, known as Hermes, B. the moon, known as Luna, C. the sun, known as Sol, D. Venus, known as Aphrodite).

Answer T or F for true or false, fill in the blank, or select the letter for the phrase that best completes the sentence.

A B	1. Carbon and sulfur are (A. metals, B. non-metals).
A B	2. The element known as brimstone is (A. sulfur, B. bismuth).
T F	3. Sulfur is one of the ingredients in gunpowder.
A B C D	4. Goodyear discovered vulcanized rubber when he (A. added carbon to sulfur, B. heated sulfur with raw rubber, C. put rubber under intense pressure, D. treated rubber with sulfuric acid).
T F	5. Goodyear made a vast fortune for his invention of vulcanized rubber.
	6. The single most important compound of sulfur is ______________ acid.
A B	7. Acid rain is due to (A. carbon, B. sulfur) being spewed into the atmosphere.
	8. The element that charcoal, coal, graphite, and diamond have in common is ______________.
A B	9. The one that is slick and can be used as a dry lubricant is (A. graphite, B. diamond).
T F	10. Synthetic diamonds are fake diamonds and have none of the properties of natural ones.
T F	11. The primary goal of alchemists was to make gold from cheap metals.
A B	12. Science in Europe (A. came to a standstill, B. made great strides) during the Middle Ages.
	13. The Middle Ages are also known as the ______________ Ages.
T F	14. The schools Robert Boyle attended taught from the latest books by Galileo and Copernicus.
A B C D	15. The invisible college was (A. a gathering of alchemists, B. a gathering of experimental scientists, C. devoted to a study of books by Ptolemy and Aristotle, D. a school for poor students).
A B	16. An element (A. can, B. cannot) be separated into simpler substances by chemical means.
A B C D	17. The Royal Society was formed to (A. buy scientific equipment, B. communicate new ideas rapidly, C. help write better science textbooks, D. teach the king about science).

Answer T or F for true or false, fill in the blank, or select the letter for the phrase that best completes the sentence.

A B 1. Henry Cavendish would be described as (A. shy, B. forward).

A B C D 2. Cavendish released the gas hydrogen by exposing metals to (A. ammonia, B. hydrochloric acid, C. carbon dioxide, D. intense heat and pressure).

A B C D 3. The name hydrogen means (A. colorless, B. lacking odor, C. lighter than air, D. water generator).

A B 4. Electricity breaks water molecules into oxygen and (A. hydrogen, B. nitrogen).

A B C D 5. If a mixture of hydrogen and oxygen are exposed to a flame (A. a violent explosion results, B. electricity is generated, C. the fire goes out, D. the mixture becomes dry).

T F 6. Hydrogen is an abundant element on earth.

A B C D 7. The reason some scientists give Joseph Priestley credit for discovering carbon dioxide is because Priestley (A. had friends who were scientists, B. had political power, C. kept his discoveries secret until the right moment, D. published his thorough studies promptly).

A B C D 8. The gas that Priestley released by heating a mercury compound was (A. carbon dioxide, B. hydrogen, C. nitrogen, D. oxygen).

A B C 9. Oxygen makes up about (A. 1/5, B. 3/4, C. all) of the atmosphere.

A B 10. The chemical activity of oxygen is (A. high, B. low).

A B 11. The one that supports combustion is (A. nitrogen, B. oxygen).

T F 12. An oxygen-acetylene cutting torch can burn under water.

A B 13. Rusting of metals is an example of (A. slow, B. fast) oxidation.

A B 14. The element in the air that prevents fires from burning too quickly is (A. oxygen, B. nitrogen).

15. Give the chemical symbols: ______ hydrogen _____ carbon ______ nitrogen ______ oxygen ______ chlorine

16. State the chemical formulas: ______ water ______ carbon dioxide ______ hydrochloric acid

Answer T or F for true or false, fill in the blank, or select the letter for the phrase that best completes the sentence.

T F 1. Static electricity was unknown to the ancient Greeks.

A B 2. Objects charged by rubbing have a (A. current, B. static) electric charge.

A B C D 3. Benjamin Franklin's kite-flying experiment proved that lightning (A. can charge a battery, B. can kill a turkey, C. helps clouds discharge rain, D. was a big discharge of static electricity).

A B C D 4. The one who discovered that frogs' legs would twitch when touched by two different metals was (A. Alessandro Volta, B. Benjamin Franklin, C. Humphry Davy, D. Luigi Galvani).

A B C D 5. Alessandro Volta built the first (A. arc lamp, B. device to make electricity by friction, C. battery to produce electric current, D. miner's safety lantern).

A B C D 6. Current electricity is due to the motion of (A. electrons, B. frog legs, C. neutrons, D. protons).

A B C D 7. The one who unwisely tasted and sniffed new chemicals was (A. Benjamin Franklin, B. Humphry Davy, C. Michael Faraday, D. Robert Wood).

A B C D 8. Davy discovered potassium by treating potash with (A. a hot arc lamp, B. heat from a large burning lens, C. carbon in a blast furnace, D. electricity from a strong battery).

9. An arc light generates light as electricity jumps across the gap between two ________________ electrodes.

A B 10. Members of the sodium family are (A. metals, B. non-metals).

A B 11. The members of the sodium family are chemically (A. active, B. inactive).

A B 12. The one used to put out a sodium fire in the laboratory is (A. water, B. dry sand).

A B 13. Compared to other metals, lithium is (A. light, B. heavy).

A B C D 14. Robert Wood used the flame test to prove that (A. lithium causes water to burn, B. lithium could be used as table salt, C. lithium is found in fireworks, D. the cook used leftovers to make stew).

Answer T or F for true or false, fill in the blank, or select the letter for the phrase that best completes the sentence.

A B 1. Sulfuric acid is an example of (A. an element, B. a compound).

A B C D 2. Dmitri Mendeleev was born in (A. China, B. Russia, C. Spain, D. the United States).

A B C D 3. The first college that Dmitri Mendeleev attended was one (A. to prepare farmers, B. to teach philosophy, C. for chemists, D. to train teachers).

A B 4. The First International Chemical Congress in Karlsruhe, Germany, was considered a (A. failure, B. success).

A B C D 5. Mendeleev believed an organized table of the elements would (A. discourage his students, B. help his students, C. show chemistry to be a difficult subject, D. make him a lot of money).

A B C D 6. Dmitri Mendeleev set about organizing the table of elements by (A. asking students to vote on their most popular element, B. collecting samples of each element, C. entering information about the elements in a computer, D. writing information about the elements on note cards).

T F 7. Dmitri Mendeleev organized his table of the elements by atomic weight and valence.

A B 8. Sodium and lithium have chemical properties that are (A. similar, B. opposite).

A B C D 9. Dmitri Mendeleev left gaps in his periodic table (A. because he couldn't remember their names, B. because he wanted to name those elements after his friends, C. for elements that had not yet been discovered, D. for elements without atomic weight).

A B C D 10. To silence critics of his table, Mendeleev (A. appealed to the government to arrest them, B. asked his friends to argue his case, C. left the country, D. predicted the properties of three missing elements).

T F 11. Only one of Mendeleev's missing elements has been found.

T F 12. Out of respect for Mendeleev, chemists refuse to make any changes to his periodic chart.

A B 13. The elements lithium, sodium, potassium, and those below it are members of the same (A. family, B. period).

Answer T or F for true or false, fill in the blank, or select the letter for the phrase that best completes the sentence.

A B C D 1. The three most important tools for making advances in chemistry were electricity, the periodic law, and the (A. law of buoyancy, B. microscope, C. spectroscope, D. telescope).

A B 2. Isaac Newton proved that white light from the sun (A. is pure light without colors, B. contains all the colors of the rainbow).

A B C D 3. Joseph von Fraunhofer (A. died in the collapse of an apartment building, B. worked as a child in a factory that glazed pottery, C. studied chemistry in a well-equipped home laboratory, D. was the son of a nobleman).

T F 4. Fraunhofer saw the lines in the spectrum while testing the quality of a prism.

T F 5. Joseph Fraunhofer invented the spectroscope in 1826 following a lecture he gave to a group of scientists.

T F 6. Robert Bunsen invented the Bunsen burner.

A B C D 7. The first element discovered by the spectroscope was (A. cesium, B. helium, C. kryptonite, D. uranium).

T F 8. Helium was discovered on the sun before it was discovered on Earth.

A B C D 9. Henry Cavendish combined oxygen with nitrogen by using (A. an electric spark, B. high heat, C. intense cold, D. great pressure).

T F 10. When Rayleigh weighed equal volumes of nitrogen freed from compounds with nitrogen separated from the atmosphere, they weighed the same.

T F 11. In 1892, Rayleigh announced that his experiment revealed a new element in the atmosphere.

A B C D 12. Argon makes up about (A. 1, B. 5, C. 78, D. 21) percent of the atmosphere of the earth.

T F 13. Argon helps make light bulbs last longer.

A B C D 14. The name helium means from (A. Helena, Montana, B. the hills, C. the sun, D. uranium).

A B C D 15. The noble gas family is also known as the (A. empire of the sun family, B. inert gases, C. Ramsay and Rayleigh families, D. strange family).

Answer T or F for true or false, fill in the blank, or select the letter for the phrase that best completes the sentence.

A B 1. John Dalton was a (A. member of the Church of England, B. Quaker).

A B 2. The word *atom* means (A. continuously divisible, B. incapable of being cut).

T F 3. According to Dalton, atoms of the same element are identical in all their properties, including their weight.

T F 4. According to Dalton, atoms of different elements differ from one another except they have the same weight.

T F 5. For many years chemists ignored John Dalton's atomic theory of matter and refused to accept it.

A B C D 6. For most of his life, John Dalton earned a living as a (A. chemist, B. minister of the gospel, C. teacher and tutor, D. weatherman).

7. The smallest objects that have the chemical properties of an element are _______________.

A B C D 8. William Crookes did scientific experimentation (A. because he enjoyed it, B. to discover a way to make synthetic gold, C. to earn money for his poor family, D. to prove the worth of English research).

A B C D 9. A radiometer (A. compares the colors of different elements, B. measures the strength of radiant energy, C. was an early form of the barometer, D. was an early form of the radio).

A B 10. The electron is (A. lighter, B. heavier) than a hydrogen atom.

A B C D 11. Most alpha particles entered the gold foil and (A. bounced out in all directions, B. caused the gold to become radioactive, C. passed straight through, D. were absorbed, never to be seen again).

A B C D 12. About 1 alpha particle in 10,000 (A. bounced out in the same direction it entered, B. caused a spark of light, C. passed through the gold foil, D. turned into gold).

A B 13. Rutherford's description of the atom is known as the (A. plum pudding, B. planetary) model of the atom.

A B C D 14. The proton has (A. a negative, B. a positive, C. a variable, D. no) electric charge.

Answer T or F for true or false, fill in the blank, or select the letter for the phrase that best completes the sentence.

A B C D 1. The plus sign, +, for the charged sodium atom, Na+, shows that it (A. carries a negative charge, B. has more protons than electrons, C. is made of a single proton, D. is too big to take part in chemical reaction).

A B C D 2. The charged sodium atom, Na+, and the charged chlorine atom, Cl-, will (A. attract one another, B. come together and destroy one another, C. produce an electric current, D. repel one another).

A B C D 3. Another name for the chlorine family is (A. acid makers, B. alkali makers, C. radioactive family, D. salt makers).

A B C D 4. The Roman word for salt is the root word for (A. acid, B. dry, C. salary, D. seawater).

5. Timbuktu is located at the edge of the _______________ Desert.

T F 6. Although salt flavors food, it has no other purpose in the diet.

A B C D 7. Henri Moissan made his laboratory equipment of platinum because (A. he had money to spare, B. it becomes solid at extremely cold temperatures, C. it reacts only slightly with fluorine, D. it is a cheap metal).

A B C 8. At room temperature bromine is a (A. gas, B. liquid, C. solid).

A B 9. Chlorinating drinking water was first tried in London in 1897 to (A. prevent tooth decay, B. stop an outbreak of typhoid fever).

A B C D 10. The acid found in the stomach that helps digestion is (A. acetic, B. hydrochloric, C. hydrofluoric, D. sulfuric) acid.

11. The purple dye containing bromine was prepared by Phoenicians who lived in the city of _______________.

A B 12. Sodium and chlorine families form compounds by (A. sharing electrons, B. electrical attraction).

Answer T or F for true or false, or select the letter for the phrase that best completes the sentence.

A B 1. Oxygen and hydrogen form the water molecule by (A. exchanging, B. sharing) electrons.

T F 2. Water is the most common liquid on earth.

A B C D 3. A diagram of a water molecule shows it as (A. a constantly shifting molecule that is never the same way twice, B. a long chain, C. an oxygen atom face with hydrogen atom ears, D. a molecule shaped like a pyramid).

A B 4. Liquid water (A. expands, B. contracts) when it changes into ice.

A B 5. Ice (A. floats, B. sinks) in water.

A B 6. Snow conducts heat (A. well, B. poorly).

A B C D 7. The Nile River continues to flow through deserts because (A. it is fed by desert springs, B. it is fed by melting snow in the mountains, C. rain falls year-round in Egypt, D. water is too heavy to evaporate).

A B C D 8. The first person to photograph a snowflake was (A. a French scientist, B. a Vermont teenager who lived on a farm, C. a Civil War photographer, D. a United States president who was an amateur scientist).

T F 9. Wilson Bentley became wealthy from his hobby of photographing snowflakes.

A B 10. In cool water, molecules move more (A. quickly, B. slowly) than in warm water.

A B C D 11. Water boils at (A. 0°C [32°F], B. 100°C [212°F], C. -161°C [-258°F], D. various temperatures depending on the phase of the moon).

A B 12. The one that boils at a hotter temperature is (A. methane, B. water).

A B C D 13. The main reason Puerto Rico has a milder climate than the Sahara Desert is that Puerto Rico (A. is an island surrounded by water, B. is closer to the equator, C. does not have a ready supply of sand, D. is nearer to the North Pole).

Answer T or F for true or false, fill in the blank, or select the letter for the phrase that best completes the sentence.

1. Carbon needs _______________ more electrons to have its electrons in the stable arrangement like the electrons of neon.

A B C D 2. Methane is also known as (A. marsh, B. mustard, C. poison, D. tear) gas.

A B C D 3. A hydrocarbon is a compound that contains only carbon and (A. chlorine, B. fluorine, C. hydrogen, D. oxygen).

A B 4. When burned with oxygen, natural gas releases carbon dioxide and (A. sulfuric acid, B. water).

T F 5. A sulfur compound is added to natural gas so a leak can be detected by the smell.

A B C D 6. The one that has been used to put people to sleep during operations is (A. carbon tetrachloride, B. chloroform, C. oxygen, D. salt).

7. The number of chlorine atoms in carbon tetrachloride is _______________.

T F 8. Carbon tetrachloride is used for dry cleaning because it is a dry powder.

T F 9. The ozone layer reduces the effect of harmful ultraviolet rays from the sun.

T F 10. Freon is used for refrigeration.

T F 11. Whenever a liquid evaporates, it warms its surroundings.

T F 12. Methanol is burned as a fuel in racing engines.

A B 13. Paraffin is an example of a (A. short, B. long) hydrocarbon chain.

A B C D 14. Teflon is an example of (A. an isobar, B. an isomer, C. a pachyderm, D. a polymer).

A B 15. Teflon is (A. sticky, B. slick).

A B C 16. Ethyl alcohol and dimethyl ether have the same (A. chemical formula, B. properties, C. structures).

A B 17. Ethyl alcohol and dimethyl ether are examples of (A. polymers, B. isomers).

A B C D 18. Petroleum means (A. floor covering, B. pet rock, C. rock oil, D. written on stone).

Answer T or F for true or false, fill in the blank, or select the letter for the phrase that best completes the sentence.

A B 1. Salt is an example of a compound that is (A. formed by living organisms, B. found in the non-living environment).

A B 2. Sugar is an example of a compound that is (A. formed by living organisms, B. found in the non-living environment).

3. The elements found in organic compounds include hydrogen, oxygen, nitrogen, and ________________.

T F 4. Chemical reactions follow a different set of rules in living things than they do in the laboratory.

T F 5. No one has ever succeeded in making organic compounds in the laboratory.

A B C D 6. Today, organic chemistry is the chemistry of (A. carbon, B. hydrogen, C. nitrogen, D. oxygen) compounds.

A B 7. Benzene is a hydrocarbon (A. chain, B. ring).

A B C D 8. The one that fights malaria is (A. aniline purple, B. Bakelite, C. benzene, D. quinine).

A B C D 9. While trying to make quinine, William Henry Perkin discovered (A. a new plastic, B. a perfume substitute, C. a synthetic dye, D. a treatment for malaria).

T F 10. Because he was a teenager, William Henry Perkin's father had to take out the patent on aniline purple.

T F 11. After he became successful, William Henry Perkin retired from chemical research.

A B C D 12. Leo Baekeland discovered the substance he called Bakelite while trying to make a substitute for (A. a dye, B. a perfume, C. eye shadow, D. shellac).

A B C D 13. When Leo Baekeland mixed carbolic acid and formaldehyde, the result was (A. a substance that clogged test tubes, B. a substance with the smell of new-mown hay, C. a thin liquid, D. an explosive gas).

T F 14. Although Bakelite was the first plastic, it was immediately replaced by better ones and proved a disappointing failure to Leo Baekeland.

A B C D 15. The word plastic means (A. capable of being shaped, B. cheap, C. phony, D. soft).

Answer T or F for true or false, fill in the blank, or select the letter for the phrase that best completes the sentence.

1. The essential element in explosives is _______________.

A B 2. The powder in the expression "keep your powder dry" was (A. diatomaceous earth, B. gunpowder).

T F 3. The rapid expansion of hot gases causes the destruction of an explosion.

T F 4. Gunpowder is smokeless.

5. Nitrocellulose is a combination of a _______________ compound with cellulose.

A B C D 6. Cellulose gives plant cells their (A. ability to do photosynthesis, B. color, C. daily supply of water, D. strong cell walls).

A B C D 7. The one that is practically pure cellulose is (A. cotton, B. diamond, C. silicon dioxide, D. protein).

A B 8. Guncotton would be described as being (A. a safe and effective explosive, B. unpredictable and capable of exploding without warning).

A B 9. Ascanio Sobrero's reaction to nitroglycerin was to (A. announce its discovery at a chemical congress, B. keep it secret).

T F 10. The Nobels sold nitroglycerin as blasting oil.

A B C D 11. Abraham Lincoln's great construction project was to connect California to the East Coast with (A. an interstate highway, B. a pony express route, C. a railroad, D. a telegraph line).

T F 12. Alfred Nobel's nitroglycerin factory in Sweden proved to be completely safe.

A B C D 13. Diatoms have cell walls of (A. cellulose, B. dynamite, C. nitroglycerin, D. silica).

A B C D 14. Blasting caps are used (A. for safe detonation of explosives, B. for small explosions, C. to contain an explosion, D. to control the direction of an explosion).

T F 15. The upsetting event that Alfred Nobel experienced was reading his own death notice in the newspaper.

T F 16. Nobel prizes can only be awarded to the citizens of Sweden.

T F 17. Nitrogen compounds are used in fertilizers.

Answer T or F for true or false, or select the letter for the phrase that best completes the sentence.

A B 1. Silicon is the (A. most abundant, B. second most abundant) element in the earth's crust.

A B C D 2. Silicon carbide was discovered while trying to make synthetic (A. diamond, B. glass, C. rubies, D. Silly Putty).

A B 3. The one that can resist sudden changes in temperature is (A. glass, B. quartz).

T F 4. Electricity passing through a quartz crystal causes it to expand and contract.

A B C D 5. Stained glass was colored (A. because church officials did not want people to look outside during church services, B. to hide the fact that glass makers could not make clear glass, C. to prevent it from melting, D. because people did not like clear glass).

A B C D 6. A gem's value is due to its (A. beauty, B. cost, C. rarity, D. all of the above).

A B C D 7. The Star of India is (A. a diamond, B. a sapphire, C. a synthetic ruby, D. an opal).

T F 8. The Hope Diamond was cut and recut until it was less than half the size of the original stone.

T F 9. A synthetic gem is a fake gem.

A B 10. The one better able to lubricate across extreme temperatures is (A. hydrocarbon motor oil, B. silicone oil).

T F 11. Silly Putty is made of long chain-like molecules that contain silicon.

A B C D 12. Silicon is an electric (A. conductor, B. insulator, C. semiconductor, D. none of the above).

A B 13. The element found in computer chips is (A. carbon, B. silicon).

Answer T or F for true or false, or select the letter for the phrase that best completes the sentence.

T F 1. Aluminum is a rare element in the earth's crust.

T F 2. Aluminum is the second most widely used metal after iron.

A B C D 3. France's Emperor Napoleon III had table sets made of (A. aluminum, B. copper, C. frozen nitrogen, D. uranium).

A B C D 4. The six-pound metallic top to the Washington Monument is made of (A. aluminum, B. gold, C. silver, D. uranium).

T F 5. The commercial production of aluminum was delayed for several years because of a court battle between Paul Héroult and Charles Martin Hall.

A B C D 6. The foil used to wrap foods is made of (A. aluminum, B. steel, C. copper, D. zinc).

T F 7. Aluminum is a better conductor of electricity than any other metal.

T F 8. Aluminum reflects heat.

T F 9. Thermite can burn under water.

A B 10. The one discovered first was the (A. planet Uranus, B. element uranium).

A B C D 11. Uranium ore is (A. galena, B. hematite, C. pitchblende, D. sulfur dioxide).

T F 12. Polonium and radium are both radioactive.

T F 13. Curie was refused a Nobel prize because she was a woman.

A B C D 14. The radioactive element used in smoke detectors is (A. americium, B. curium, C. radium, D. uranium).

T F 15. Some elements beyond uranium were named after states and cities.

Write the matching letter in the blank provided.

1.________ helped found the Royal Society. He also defined an element.

2.________ was an eccentric English chemist who discovered hydrogen.

3.________ was a Frenchman who burned a diamond. He stated the law of conservation of matter.

4.________ stated the atomic theory of matter.

5.________ used electricity to free sodium and other elements from their ores.

6.________ suggested the use of chemical symbols for elements.

7.________ was Davy's assistant who discovered benzene.

8.________ was a French chemist who became a medical researcher.

9.________ made the first periodic table of the elements.

10.________ discovered the family of inert gases.

a. Jöns Jakob Berzelius

b. Robert Boyle

c. Henry Cavendish

d. John Dalton

e. Humphry Davy

f. Michael Faraday

g. Antoine Laurent Lavoisier

h. Dmitri Ivanovich Mendeleev

i. Louis Pasteur

j. William Ramsay

Quizzes and Tests Section

	Exploring Mathematics Concepts & Comprehension	Quiz 1	Scope: Chapters 1–4	Total score: ____of 100	Name

Matching (2 Points Each Question)

1. _____ day — a. due to the tilt of the earth's axis, equal to three months
2. _____ week — b. earth revolves around the sun once
3. _____ month — c. earth rotates on its axis once
4. _____ season — d. moon revolves around the earth once
5. _____ year — e. seven days

Fill-in-the-Blank Questions (4 Points Each Question)

6. The length of a mile in feet is _______________.

7. "A pint is a _______________ the world around."

Multiple Choice Questions (4 Points Each Question)

8. The first calendar with a leap day every four years was the one
   A. authorized by Julius Caesar
   B. used by the American colonies after 1752
   C. used by the Babylonians
   D. used by the Egyptians

9. The inventors of the hourglass were the
   A. Babylonians
   B. British Navy
   C. Egyptians
   D. Romans

10. Military time has hours numbered from 0000 to
    A. 0400
    B. 1200
    C. 2400
    D. 3600

11. Time zones were introduced when it became common to travel by
    A. airplanes
    B. ox carts
    C. ships
    D. trains

12. NASA's Climate Orbiter to Mars failed because
    A. American and French engineers did not communicate with one another
    B. engineers used two different measures of force
    C. fuel had been measured improperly
    D. the spacecraft weighed too much

13. A scruple was a standard of weight for measuring
    A. barley
    B. diamonds
    C. drugs
    D. potatoes

14. Most early measures of distance were based on
    A. animal strides
    B. the human body
    C. parts of ships
    D. Roman military terms

15. Until 1960, the meter was considered
A. 1,640,763.73 wavelengths of krypton gas
B. 1/10,000,000 of the distance from the equator to the North Pole
C. the distance between two scratch marks on a metal rod
D. the distance light travels in 1/299,792,458 of a second

16. Daniel Fahrenheit set the boiling temperature of water on his thermometer at
A. 0 degrees B. 32 degrees
C. 100 degrees D. 212 degrees

Multiple Answer Questions (3 Points Each Answer – 18 Points Total)

17. Using the Babylonian calendar of 360 days in a year, how many days are in one-third of a year; one-fifth of a year; one-twentieth of a year; one-sixtieth of a year?

a. b. c. d.

18. A hand is four inches. How tall is a horse in inches that is 15 hands tall? How tall in feet?

a. b.

Short Answer Questions (4 Points Each Question)

19. What is the main reason to have leap days?

20. Assume that the first four-hour watch began at midnight. What time would it be at five bells on the second watch?

21. At 4:00 p.m., a family on vacation drives from Mountain Standard Time into Central Standard Time. Should their watches be set one hour earlier to 3:00 p.m. or one hour later to 5:00 p.m.?

22. The tallest mountain on earth is Mt. Everest. Its summit is 29,035 feet above sea level. How high is the mountain in miles?

Applied Learning Activities (2 Points Each Answer)

23. Feel your pulse at the wrist and count the number of beats in a minute. Calculate the number of times your heart beats in a day.

24. Change your weight from pounds to ounces.

Choose the larger:
25. A. foot B. yard
26. A. fathom B. yard
27. A. nautical mile B. statute mile
28. A. cup B. quart

Exploring Mathematics Concepts & Comprehension	Quiz 2	Scope: Chapters 5–8	Total score: ____of 100	Name

Matching (2 Points Each Question)

1. _____ circle — a. a polygon with five sides
2. _____ pentagon — b. a rectangle with four equal sides
3. _____ rectangle — c. a polygon with three sides and one right angle
4. _____ right triangle — d. a quadrilateral with opposite sides parallel and equal in length
5. _____ square — e. is not a polygon

6. _____ 2L + 2W — a. area of a circle
7. _____ 4S — b. area of a rectangle
8. _____ Ah — c. area of a square
9. _____ L x W — d. perimeter of a rectangle
10. _____ pr2 — e. perimeter of a square
11. _____ S2 — f. volume

12. _____ Archimedes — a. discovered that the sum of the 3 angles of any triangle is 180 degrees
13. _____ Eucli — b. used ratios to find the heights of buildings
14. _____ Johannes Kepler — c. proved planets follow elliptical orbits
15. _____ Pythagoras — d. wrote *Elements of Geometry*
16. _____ Thales — e. ancient Greek who worked out a way to show large numbers that he called myriads

17. _____ circle — a. a mirror of this shape will focus sunlight
18. _____ ellipse — b. all points are the same distance from the center
19. _____ parabola — c. the first part of the name means over or beyond
20. _____ hyperbola — d. the orbit of Halley's comet is of this shape

21. _____ 1881, 121, 1001 — a. Fibonacci numbers
22. _____ 2, 3, 5, 7, 11, 13 . . . — b. palindromes
23. _____ 1, 1, 2, 3, 5, 8, 13 . . . — c. prime numbers
24. _____ 1, 4, 9, 16, 25 . . . — d. square numbers
25. _____ 1, 3, 6, 10, 15 . . . — e. triangular numbers

Fill-in-the-Blank Questions (2 Points Each Answer)

26. The sum of the ______________ of the legs of a right triangle are equal to the ______________ of the hypotenuse.

27. The next Fibonacci number after 89 and 144 is ______________.

Multiple Choice Questions (4 Points Each)

28. The Egyptian knotted rope was used to measure out
    - A. a pyramid with sloping sides
    - B. a rectangle with parallel sides
    - C. a silo of a fixed height
    - D. a triangle with a right angle

29. Doubling the length, width, and height of a box gives it a volume
    - A. twice as great
    - B. three times as great
    - C. six times as great
    - D. eight times as great

30. A whispering gallery has a shape like
    - A. a circle
    - B. a hyperbola
    - C. a parabola
    - D. an ellipse

31. The study of the properties of whole numbers is called
    - A. algebra
    - B. geometry
    - C. number theory
    - D. real analysis

True and False (2 Points Each)

32. T F Mathematics is sometimes called the ruler of science.

33. T F Isaac Newton introduced the use of place value and the numeral 0 to Europe.

34. T F The prefix *bi* means one-half.

35. T F The word billion has the same meaning in England as in the United States.

36. T F Of the prefixes giga, mega, and tera, the one that has the greatest value is mega.

Short Answer Questions (4 Points)

37. A room is 10 feet wide and 14 feet long. How many square tiles, one foot on a side, would be needed to completely cover the room?

Applied Learning Activity (14 Points Total: 2 Points each Square)

38. Draw a Fibonacci spiral.

Exploring Mathematics Concepts & Comprehension	Quiz 3	Scope: Chapters 9–11	Total score: ____of 100	Name

Matching (2 Points Each Question)

1. ____ 1, 2, 3, 4, 5
2. ____ 0, 1, 2, 3, 4, 5
3. ____ 2, 4, 6, 8, 10
4. ____ 1, 3, 5, 7, 9
5. ____ -3, -2, -1, 0, +1, +2, +3
6. ____1/1, 1/2, 3/4, 2/3
7. ____ SR2, p, (1 + SR5)/2

a. counting numbers
b. even numbers
c. integers
d. irrational numbers
e. odd numbers
f. rational numbers
g. whole numbers

Use the figure to match the equations (questions 8–11) and the statements (questions 12–15)

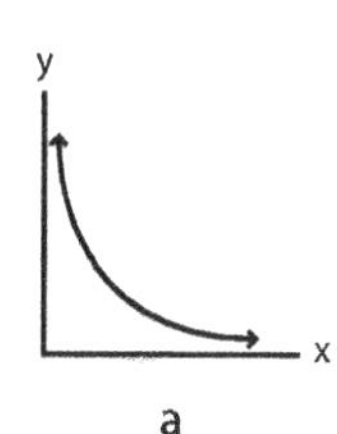

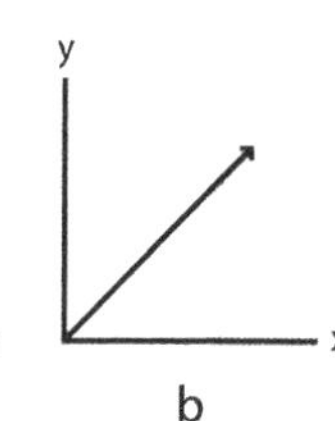

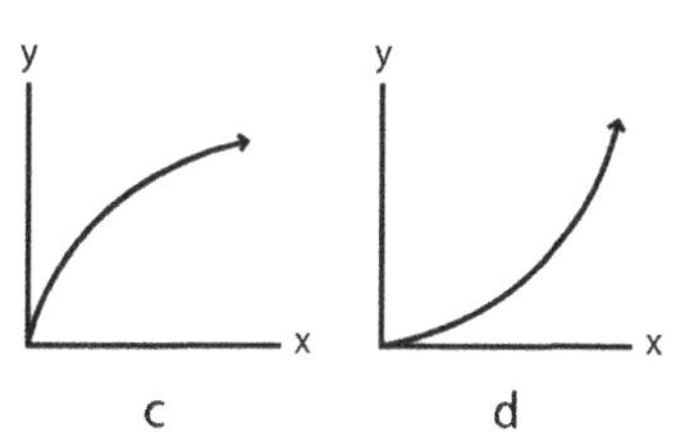

Equations:

8. ____ y = kx 9. ____ y = kx2 10. ____ y = kSRx 11. ____ y = k / x

Statements:

12. ____ The length a spring stretches (y-axis) is directly proportional to the force pulling on the spring (x-axis).
13. ____ The distance an object falls (y-axis) under the influence of gravity is directly proportional to the square of the time it has fallen (x-axis).
14. ____ The volume of a gas (y-axis) is inversely proportional to the pressure acting on the gas (x-axis).
15. ____ The period of a pendulum (y-axis) is directly proportional to the square root of the length of the pendulum (x-axis).

16. ____ Andrew Wiles
17. ____ Blaise Pascal
18. ____ Isaac Newton
19. ____ Leonhard Euler
20. ____ Pierre de Fermat

a. discovered how to calculate the coefficients of a binomial raised to a power
b. he called his triangle an arithmetic triangle
c. his last theorem was solved in 1995
d. solved the Konigsberg bridge problem
e. proved that xn + yn = zn has no solution with whole numbers except for n = 2

Multiple Choice Questions (4 Points Each Question)

21. Before the invention of calculators, shares were used to reduce the necessity of doing
    A. addition
    B. subtraction
    C. multiplication
    D. division

22. The American colonies divided the real, a Spanish coin, into how many pieces?
    A. 2 B. 4 C. 8 D. 12

23. Two bits is equal to how many cents?
    A. 121/2 B. 25 C. 50 D. 100

24. A common fraction can be changed into a decimal by dividing the numerator by the
    A. denominator
    B. greatest common factor
    C. least common multiple
    D. remainder

25. The square root of two, is an example of
    A. a common fraction
    B. an irrational number
    C. a rational number
    D. a terminating decimal

26. The digits of the square root of two, $\sqrt{2}$, when expressed as a decimal
    A. do not repeat
    B. do not terminate
    C. do not form a pattern
    D. all of the above

27. Discovering the value of x when y is equal to zero is called
    A. modernizing
    B. normalizing
    C. solving
    D. zeroing the equation

28. The 5 in the equation $5x + 2 = 0$ is called
    A. a coefficient
    B. a constant
    C. an equalizer
    D. a variable

29. In the 1600s, the French mathematician Rene Descartes discovered a way to combine algebra with
    A. computer programming
    B. geometry
    C. number theory
    D. physics

30. The problem that a computer helped solve was the
    A. bell peal problem
    B. binomial theorem
    C. four-color map problem
    D. Konigsberg bridge problem

31. The expression 3! is read as "three
    A. combinations"
    B. factorial"
    C. permutations"
    D. probabilities"

32. The value of 5! is
    A. 24 B. 25 C. 120 D. 125

Applied Learning Activities (4 Points)

33. The state of Missouri has license plates with three letters followed by three digits. How many license plates are possible?

(8 Points Total; 2 Points Each Answer)

34. Four Corners is the only point in the United States where four states touch a single point. Name the states.

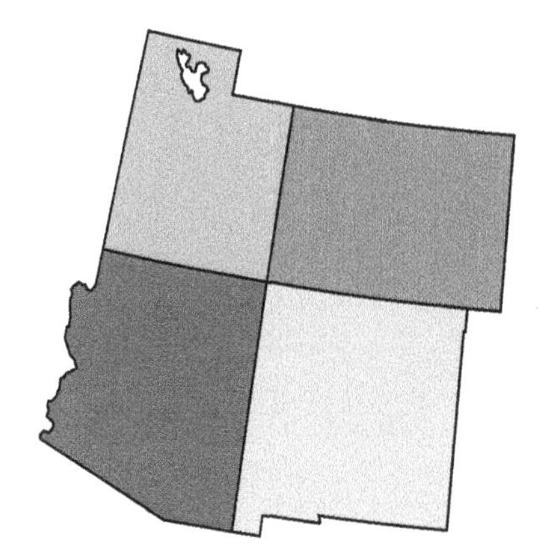

Exploring Mathematics Concepts & Comprehension	Quiz 4	Scope: Chapters 12–14	Total score: ____of 100	Name

Matching (2 Points Each Question)

1. ____ Augusta Ada Byron, Lady Lovelace
2. ____ Howard H. Aiken
3. ____ Charles Babbage
4. ____ Herman Hollerith
5. ____ Johannes Kepler
6. ____ Gottfried Leibnitz
7. ____ John Napier
8. ____ Blaise Pascal

a. built a calculator called the Step Reckoner
b. built a calculator to help his father, a tax collector
c. built the first general-purpose calculating machine
d. built the difference engine
e. invented logarithms
f. invented tabulating machines used in 1890 census
g. spent six years calculating the orbit of Mars
h. wrote the first computer program

Fill-in-the-Blank Questions (2 Points Each Answer)

9. The expression log103 = 0.477 is read as "The logarithm of the number __________ in base __________ is __________."

10. Base 10 uses the digits 0 through ______________.

Multiple Choice Questions (5 Points Each Question)

11. Any number raised to the zero power is
A. 0 B. 1 C. 2 D. undefined

12. The number 5,280 changed to standard notation is
A. .5280 x 101 B. 5,280 x 103 C. 5.28 x 102 D. 5.28 x 103

13. In the early days of computers, input was mainly by 12
A. colored ribbons B. punched cards
C. spoken commands D. switches and relays

14. The letters RAM stand for
A. random access memory B. reasonably accurate member
C. recent abacus modification D. Robert A. Morley

15. In computer usage, a single position for a binary digit is called
A. bit B. byte C. kilo D. pixel

16. In personal computers, a byte of data is made of how many bits?
A. one B. two C. eight D. ten

17. In 1861, James Clerk Maxwell made a color photograph using
A. computer enhancement B. color ink drops sprayed on paper
C. polarized light D. the three colors of red, green, and blue

18. Video images can be compressed by
    A. converting black and white images to color images
    B. having reporters avoid standing in front of a blue sky
    C. transmitting all pixels that are the same as the previous one
    D. transmitting only those pixels that are different from the previous one

19. Moore's law states that computers double in power every
    A. days B. months C. decades D. years

20. A computer with components put farther apart will run more slowly because
    A. electric signals can go no faster than the speed of light
    B. larger components must be made of less costly materials
    C. of resistance in the wires
    D. the electrons get lost

Short Answer Questions (4 Points Each Question)

21. In this list, which one is considered the "heart" of a computer: input, control program, memory, central processing unit, output?

22. The Constitution of the United States has 4,609 words and 26,747 characters. At the rate of 7,000 bytes per second, how long would it take a computer to download the Constitution of the United States as an uncompressed text file?

Applied Learning Activities (6 Points Each Puzzle/Riddle)

23. Samson's Riddle — The Bible has puzzles such as Samson's riddle in Judges 14:14: He replied, "Out of the eater, something to eat; out of the strong, something sweet." Hint: You can find the answer in Judges 14:8.

24. On the road to St. Ives — Try to solve this people-on-the-road puzzle that was turned into English children's rhyme:

    As I was going to St. Ives I met a man with seven wives;
    Every wife had seven sacks; Every sack had seven cats;
    Every cat had seven kits [kittens]; Kits, cats, sacks, and wives,
    How many were going to St. Ives?

    Can you figure out the answer to the riddle?

25. River Crossing — A canoeist must cross a river with three things, but his canoe can hold only one thing at a time. How can the canoeist get a wolf, goat, and carrots across a river? If left alone, the wolf would eat the goat, and the goat would eat the carrots.

Exploring Mathematics Concepts & Comprehension	Test 1	Scope: Chapters 1–14	Total score: ____of 100	Name

Matching (1 Points Each Question)

1. ____ day
2. ____ week
3. ____ month
4. ____ season
5. ____ year

a. due to the tilt of the earth's axis, equal to three months
b. earth revolves around the sun once
c. earth rotates on its axis once
d. moon revolves around the earth once
e. seven days

6. ____ circle
7. ____ ellipse
8. ____ parabola
9. ____ hyperbola

a. a mirror of this shape will focus sunlight
b. all points are the same distance from the center
c. the first part of the name means over or beyond
d. the orbit of Halley's comet is of this shape

10. ____ 1, 2, 3, 4, 5
11. ____ 0, 1, 2, 3, 4, 5
12. ____ 2, 4, 6, 8, 10
13. ____ 1, 3, 5, 7, 9
14. ____ -3, -2, -1, 0, +1, +2, +3
15. ____1/1, 1/2, 3/4, 2/3
16. ____ SR2, p, (1 + SR5)/2

a. counting numbers
b. even numbers
c. integers
d. irrational numbers
e. odd numbers
f. rational numbers
g. whole numbers

17. ____ Augusta Ada Byron, Lady Lovelace
18. ____ Howard H. Aiken
19. ____ Charles Babbage
20. ____ Herman Hollerith
21. ____ Johannes Kepler
22. ____ Gottfried Leibnitz
23. ____ John Napier
24. ____ Blaise Pascal

a. built a calculator called the Step Reckoner
b. built a calculator to help his father, a tax collector
c. built the first general-purpose calculating machine
d. built the difference engine
e. invented logarithms
f. invented tabulating machines used in 1890 census
g. spent six years calculating the orbit of Mars
h. wrote the first computer program

Fill-in-the-Blank Questions (2 Points Each Answer)

25. The length of a mile in feet is ______________.

26. "A pint is a ______________ the world around."

27. The sum of the ______________ of the legs of a right triangle are equal to the ______________ of the hypotenuse.

28. The next Fibonacci number after 89 and 144 is ______________.

29. The expression log103 = 0.477 is read as "The logarithm of the number __________ in base __________ is __________."

30. Base 10 uses the digits 0 through ______________.

Multiple Choice Questions (2 Points Each Question)

31. NASA's Climate Orbiter to Mars failed because
    A. American and French engineers did not communicate with one another
    B. engineers used two different measures of force
    C. fuel had been measured improperly
    D. the spacecraft weighed too much

32. A scruple was a standard of weight for measuring
    A. barley C. drugs B. diamonds D. potatoes

33. Until 1960, the meter was considered
    A. 1,640,763.73 wavelengths of krypton gas
    B. 1/10,000,000 of the distance from the equator to the North Pole
    C. the distance between two scratch marks on a metal rod
    D. the distance light travels in 1/299,792,458 of a second

34. Daniel Fahrenheit set the boiling temperature of water on his thermometer at
    A. 0 degrees C. 100 degrees B. 32 degrees D. 212 degrees

35. The Egyptian knotted rope was used to measure out
    A. a pyramid with sloping sides B. a rectangle with parallel sides
    C. a silo of a fixed height D. a triangle with a right angle

36. Doubling the length, width, and height of a box gives it a volume
    A. twice as great B. three times as great
    C. six times as great D. eight times as great

37. A whispering gallery has a shape like
    A. a circle B. a hyperbola C. a parabola D. an ellipse

38. The study of the properties of whole numbers is called
    A. algebra B. geometry C. number theory D. real analysis

39. A common fraction can be changed into a decimal by dividing the numerator by the
    A. denominator B. greatest common factor
    C. least common multiple D. remainder

40. The square root of two, is an example of

A. a common fraction
B. an irrational number
C. a rational number
D. a terminating decimal

41. In the 1600s, the French mathematician Rene Descartes discovered a way to combine algebra with

A. computer programming
B. geometry
C. number theory
D. physics

42. The problem that a computer helped solve was the

A. bell peal problem
B. binomial theorem
C. four-color map problem
D. Konigsberg bridge problem

43. In 1861, James Clerk Maxwell made a color photograph using

A. computer enhancement
B. color ink drops sprayed on paper
C. polarized light
D. the three colors of red, green, and blue

44. Video images can be compressed by

A. converting black and white images to color images
B. having reporters avoid standing in front of a blue sky
C. transmitting all pixels that are the same as the previous one
D. transmitting only those pixels that are different from the previous one

45. Moore's law states that computers double in power every 18

A. days B. months C. decades D. years

46. A computer with components put farther apart will run more slowly because

A. electric signals can go no faster than the speed of light
B. larger components must be made of less costly materials
C. of resistance in the wires
D. the electrons get lost

Multiple Answer Question (3 Points Each Answer)

47. A hand is four inches. How tall is a horse in inches that is 15 hands tall? How tall in feet?

a. b.

Short Answer Questions (2 Points Each Question)

48. What is the main reason to have leap days?

49. At 4:00 p.m., a family on vacation drives from Mountain Standard Time into Central Standard Time. Should their watches be set one hour earlier to 3:00 p.m. or one hour later to 5:00 p.m.?

50. The tallest mountain on earth is Mt. Everest. Its summit is 29,035 feet above sea level. How high is the mountain in miles?

51. A room is 10 feet wide and 14 feet long. How many square tiles, one foot on a side, would be needed to completely cover the room?

Applied Learning Activities (6 Points Each Puzzle)

52. Sock Puzzle — Because of an electrical power failure, a boy must get dressed in a dark bedroom. His sock drawer has 10 blue socks and 10 black socks, but in the darkness he cannot tell them apart. He dresses anyway. He reaches into the drawer to grab spare socks so he can change into matching colors later. How many should he take to be certain he has a matching pair?

53. Durer's Number Square — Make a number square by using the digits 1 through 9 in a three by three square. Each of the rows, columns, and diagonals should add to the same number. Eight different squares are possible.

Exploring Physics Concepts & Comprehension	Quiz 1	Scope: Chapters 1–4	Total score: ____of 100	Name

Matching (2 Points Each)

1. ____ first law of motion — a. a = f/m
2. ____ second law of motion — b. f = m x a
3. ____ third law of motion — c. $f_{ab} = -f_{ba}$
4. ____ force equation — d. I = f x t
5. ____ definition of impulse — e. If f = 0 then a = 0
6. ____ definition of momentum — f. p = m x v

Fill-in-the-Blank Questions (4 Points Each Answer)

7. To calculate speed, divide distance by ______________.

8. Suppose a canoeist takes 70 days to paddle the entire length of the Mississippi River, a distance of 3,710 miles. The canoeist's average speed in miles per day is ______________.

9. On the moon, the acceleration due to gravity is 5.3 ft/sec_2 rather than 32 ft/sec_2. If an object fell six seconds before hitting the ground, it strikes the ground with a speed of ______________ ft/sec. (Hint: Use the final velocity equation.)

10. Momentum is the mass of an object times its ______________.

11. Force of gravitational attraction between two objects is directly proportional to the ______________ of their masses and inversely proportional to the ______________of the distance separating them.

12. The Grand Canyon is about one mile deep, and the most popular trail out of the canyon is nine miles long; the mechanical advantage of the trail is ______________.

Multiple Choice Questions (4 Points Each)

13. Physics is the science that explores how energy acts on
    A. heat, B. light, C. matter, D. sound.

14. A feather and lump of lead will fall at the same speed in
    A. a high speed wind tunnel, B. the atmosphere, C. a vacuum, D. water

15. To study the motion of falling objects, Galileo
    A. beat them into cubes
    B. dropped them from a high tower
    C. pushed them from a cliff
    D. rolled them down a ramp

16. Acceleration is found by dividing the
    A. average velocity
    B. distance
    C. gravity
    D. change in speed by the change in time.

17. Inertia is a property of matter that resists changing its
A. electric charge, B. mass, C. momentum, D. velocity

18. Kepler proved that planets traveled in orbits that were
A. circular, B. elliptical, C. parabolic, D. straight-line

19. The Greek who said, "Give me a place to stand and a long enough lever, and I can move the world" was
A. Archimedes, B. Aristotle, C. Eratosthenes, D. Ptolemy

20. The tab on a soft drink can is an example of
A. an inclined plane, B. a lever, C. a pulley, D. a wheel and axle

Short Answer Questions (4 Points Each Question)

21. An ordinary passenger car can accelerate to 60 miles per hour in about eight seconds. What is the car's acceleration?

22. State the second law of motion.

Applied Learning Activities (2 Points Each Answer – 20 Points Total)

23. Label the four forces acting on an airplane in flight.

24. Label the fulcrum, load, and effort points on the seesaw.

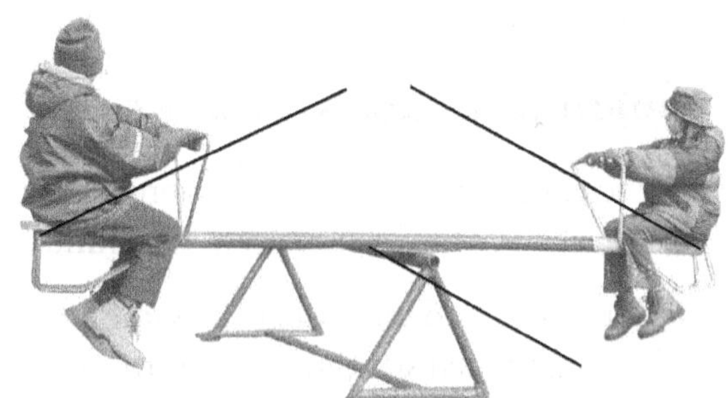

25. Label the fulcrum, load, and effort points on the nutcracker.

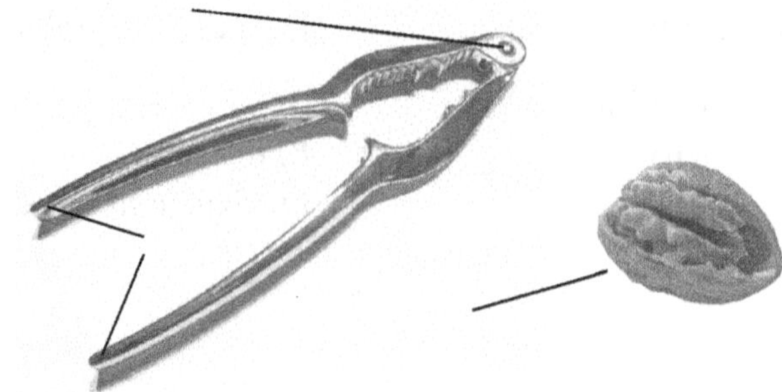

	Exploring Physics Concepts & Comprehension	Quiz 2	Scope: Chapters 5–7	Total score: ____of 100	Name

Matching (2 Points Each)

1. ____ Archimedes' principle of buoyancy
2. ____ Boyle's law
3. ____ Ideal gas law
4. ____ Bernoulli's principle

   a. Pressure times volume of any gas divided by the temperature is a constant.

   b. The lifting force acting on a solid object immersed in water is equal to the weight of the water shoved aside by the object.

   c. The velocity of a fluid and its pressure are inversely related.

   d. The volume of a gas is inversely proportional to the pressure.

Fill-in-the-Blank Questions (4 Points Each Answer)

5. Work transfers ________________ from one place to another.

6. James Prescott Joule found how mechanical energy due to motion compares to ________________ energy.

7. The three factors that determine the heat contained in an object are type of substance, mass, and ________________.

8. The maximum efficiency possible for a machine that produces energy from the difference of ocean water at 18°C at the surface and 1°C at depth is ________________.

Multiple Choice Questions (4 Points Each)

9. The equation E = f x d is used to find
   A. efficiency B. mechanical advantage
   C. momentum D. work

10. Foot-pounds (English system) and joules (metric system) both measure
    A. force B. mass C. power D. work

11. Heat is a type of A. energy B. force C. matter D. temperature

12. The two most common substances used in thermometers are colored alcohol and
    A. cooking oil B. ethylene glycol C. mercury D. molten salt

13. The scientist who discovered that pure water has a fixed boiling and freezing temperature was
    A. Anders Celsius B. Antoine Lavoisier C. Daniel Fahrenheit D. John Dalton

14. Density is equal to mass divided by
    A. area B. pressure C. volume D. weight

15. The English and metric system units for measuring power are
A. calorie and joule B. pound and newton C. horsepower and watt

16. Almost every time that energy changes form, the amount of what kind of energy increases?
A. heat B. kinetic C. potential

17. Heat moving from one end of a metal fireplace poker to the other end is an example of heat transfer by
A. conduction B. convection C. radiation

18. A sea breeze is set in motion because of A. conduction B. convection C. radiation

19. The rate of diffusion of a gas is inversely proportional to the ______ of its molecular weight.
A. square B. square root C. sum

True and False (2 Points Each)

20. T F Energy is a term that has been in use for more than 2,000 years.

21. T F Energy can be changed from one form to another.

22. T F Pushing against a desk that does not move is an example of work.

23. T F Doubling mass of a moving object doubles its kinetic energy.

24. T F Doubling velocity of a moving object doubles its kinetic energy.

25. T F Scientists are unable to measure temperatures greater than 1,700°F.

26. T F Heat is the motion of atoms and molecules.

27. T F Heat is transferred from the sun to earth by radiation.

28. T F A steam engine works because heat flows from a hot region to a cold region.

29. T F Moving heat energy in a direction opposite to its normal flow requires work.

30. T F A rubber band is elastic because it will stretch.

31. T F Steel is highly elastic.

32. T F The pressure of a liquid acts equally in all directions.

Applied Learning Activity (6 Points)

33. What does this demonstration measure?

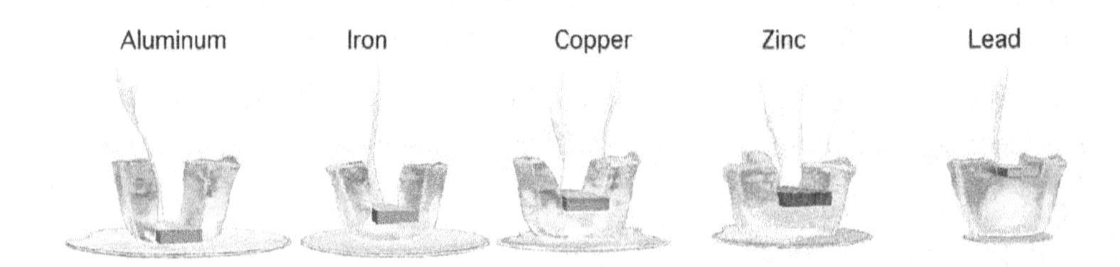

	Exploring Physics Concepts & Comprehension	Quiz 3	Scope: Chapters 8–10	Total score: ____of 100	Name

Matching (2 Points Each Question)

1. ____ brings light to a focus.
2. ____ controls the amount of light that enters the eye.
3. ____ is the opening through which light enters the eye.
4. ____ adjusts light to the best focus.
5. ____ is a surface of light sensitive nerves.
6. ____ carries information from the eye to the brain.
7. ____ is sensitive to light but cannot see color.
8. ____ is sensitive to light and can distinguish color.

a. Cones
b. Cornea
c. Iris
d. Lens
e. Optic nerve
f. Pupil
g. Retina
h. Rods

Fill-in-the-Blank Questions (2 Points Each Answer)

9. The pitch of a string on a stringed instrument depends on the length, thickness, and ________________ of the string.
10. The three properties of a sound are frequency, intensity, and ________________.
11. The eye has cones that can detect red, green, and ________________ light.
12. Coulomb's law of static electric force is very similar to Newton's law of gravity, but with ________________ replacing mass.
13. To reduce the heating effect of electricity in wires, the current is reduced but the ________________ is increased.

Underline the Correct Answer (2 Points Each Answer)

14. The (A. highest, B. lowest) pitch an object can make is known as its natural or fundamental frequency.
15. High frequency, ultrasonic sounds reflect (A. better, B. worse) from small objects than low frequency sounds.
16. Sound waves travel at the (A. same speed, B. different speeds) in air depending on their source.
17. If pitch increases, then source and observer must be moving (A. toward, B. away from) one another.
18. The observation that light bounces from a mirror at the same angle at which it enters is known as the law of (A. reflection, B. refraction).
19. The image behind a flat mirror is a (A. real, B. virtual) image.
20. Most modern large telescopes use a (A. lens, B. mirror) to collect light and bring it to a focus.
21. A lens thicker in the middle than at the edges is (A. convex, B. concave).

22. The speed of light is (A. faster, B. slower) in water than in air.

23. The bending of the sun's rays at sunset is an example of (A. refraction, B. reflection).

24. The one that moves more freely is the (A. electron, B. proton).

25. The stronger force is (A. electrostatic, B. gravity).

26. Glass is an example of a (A. conductor, B. nonconductor).

Multiple Choice Questions (2 Points Each Question)

27. The distance along a wave, including crest and trough, is its
A. axis, B. frequency, C. velocity, D. wavelength

28. The frequency of a sound is known as its A. amplitude, B. color, C. pitch, D. velocity

29. The study of sound is known as A. acoustics, B. astronomy, C. mechanics, D. thermodynamics

30. The loudness of sound is measured in A. candles, B. decibels, C. joules, D. watts

31. The force that pushes electrons around a circuit is A. resistance, B. charge, C. current, D. voltage

32. The ohm is a unit for measuring A. current, B. power, C. resistance, D. voltage

Short Answer Question (6 Points)

33. State Ohm's law:

True and False (2 Points Each)

34. T F As sound waves spread out, they grow weaker by the square of the distance.

35. T F Sunlight is a mixture of all the colors of the rainbow.

36. T F The frequency of light is its brightness.

37. T F Thales of Melitus discovered that amber could be given a charge of static electricity.

38. T F An object with a positive charge has more protons than electrons.

39. T F All metals conduct electricity.

40. T F No practical use has been found for battery-powered vehicles.

Applied Learning Activity (2 Points Each)

Identify the sources for each decibel level:

Decibels

41. — 8

42. — 10–20

43. — 20–30

44. — 40–50

45. — 50–60

46. — 70–80

47. — 90–100

48. — 110

a. Heavy street traffic

b. Automobile

c. Thunder

d. Whisper

e. Ordinary conversation

f. Average home sounds (such as the humming of a refrigerator)

g. Rustling of leaves

h. Jack hammer

	Exploring Physics / Concepts & Comprehension	Quiz 4	Scope: Chapters 11–14	Total score: ____of 100	Name

Matching (3 Points Each Question)

1. _____ Niels Bohr
2. _____ Louis de Borglie
3. _____ Max Planck
4. _____ Werner Heisenberg

a. developed a model of the atom and electron orbits
b. explained black body radiation by using energy quanta
c. developed the uncertainty principle
d. proposed matter waves and calculated their wavelengths

Fill-in-the-Blank Questions (4 Points)

5. The Heisenberg uncertainty principle states that the precise position, mass, and ________________ cannot be determined exactly.

Multiple Choice Questions (4 Points Each Question)

6. The advantage of an electromagnet is that it
   A. can be turned on and off
   B. does not follow the inverse square law
   C. can both attract and repel iron
   D. takes less electricity to operate than a natural magnet

7. Faraday succeeded in showing a connection between
   A. chemistry and electricity
   B. electricity and magnetism
   C. magnetism and light
   D. all of the above

8. The scientist who developed four equations that summarized electromagnetism was
   A. Albert Einstein
   B. Isaac Newton
   C. James Clerk Maxwell
   D. Michael Faraday

9. The first scientist to generate electromagnetic waves was
   A. Arthur Compton
   B. Guglielmo Marconi
   C. Michael Faraday
   D. Rudolf Hertz

10. The M in AM and FM stands for
    A. Marconi, B. Maxwell, C. modulation, D. momentum

11. The period 1895–1905 is known as
    A. the Aristotle period
    B. the atomic age
    C. the first scientific revolution
    D. the second scientific revolution

12. Albert Einstein won the Nobel Prize in physics because of his research papers about
    A. Brownian motion
    B. the equation E = mc2
    C. photoelectric effect
    D. special theory of relativity

13. The one that is not made of smaller particles is
    A. an electron
    B. a hydrogen atom
    C. a neutron
    D. a proton

14. Enrico Fermi discovered that the best particles to cause the break-up of a nucleus were
    A. electrons
    B. helium nuclei
    C. neutrons
    D. protons

15. The first person to state that uranium could undergo fission and produce a self-sustaining chain reaction was

A. Albert Einstein B. Enrico Fermi C. Franklin D. Roosevelt D. Lise Meiter

16. The purpose of a moderator is to

A. absorb neutrons B. cool the reactor C. generate heat D. slow neutrons

17. The first particle shown to have a wave nature was

A. an alpha particle B. an electron C. a gamma ray D. a proton

Underline the Correct Answer (2 Points Each Answer)

18. The one that is more difficult to magnetize is (A. soft iron, B. steel).

19. When magnetic domains become jumbled, magnetism is (A. lost, B. strengthened).

20. The (A. AM, B. FM) radio band is prone to electrical interference.

21. The splitting of an atom into smaller parts is nuclear (A. fission, B. fusion).

22. The total mass after a nuclear reaction is (A. less, B. more) than the total mass before the reaction.

23. In the black box experiment, the amount of ultraviolet light was (A. far less, B. many times greater) than predicted.

24. An ultraviolet quantum has (A. more, B. less) energy than an infrared quantum.

25. The one with a wave long enough to be detected is (A. an electron, B. a baseball).

26. The ground state, or lowest orbit of an electron, corresponds to its (A. fundamental frequency, B. highest overtone).

Applied Learning Activities (2 Points Each Answer)

27. Write the numbers 1 to 4 in the blanks to rank these waves in order from lowest frequency (longest wavelength) to highest frequency (shortest wavelength):

_____ blue visible light

_____ AM radio waves

_____ X rays

_____ infrared light

28. Label the elementary particles of an atom:

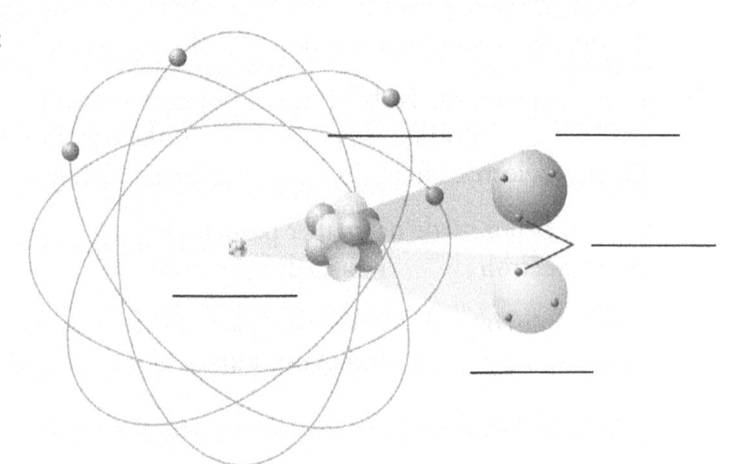

Exploring Physics / Concepts & Comprehension	Test 1	Scope: Chapters 1–14	Total score: ____of 100	Name

Matching (1 Point Each Question)

1. ____ first law of motion — a. $a = f/m$
2. ____ second law of motion — b. $f = m \times a$
3. ____ third law of motion — c. $f_{ab} = -f_{ba}$
4. ____ force equation — d. $I = f \times t$
5. ____ definition of impulse — e. If $f = 0$ then $a = 0$
6. ____ definition of momentum — f. $p = m \times v$

7. ____ Archimedes' principle of buoyancy
8. ____ Boyle's law
9. ____ Ideal gas law
10. ____ Bernoulli's principle

a. Pressure times volume of any gas divided by the temperature is a constant.

b. The lifting force acting on a solid object immersed in water is equal to the weight of the water shoved aside by the object.

c. The velocity of a fluid and its pressure are inversely related.

d. The volume of a gas is inversely proportional to the pressure.

11. ____ brings light to a focus. — a. Cones
12. ____ controls the amount of light that enters the eye. — b. Cornea
13. ____ is the opening through which light enters the eye. — c. Iris
14. ____ adjusts light to the best focus. — d. Lens
15. ____ is a surface of light sensitive nerves. — e. Optic nerve
16. ____ carries information from the eye to brain. — f. Pupil
17. ____ is sensitive to light but cannot see color. — g. Retina
18. ____ is sensitive to light and can distinguish color. — h. Rods

19. ____ Niels Bohr — a. developed a model of the atom and electron orbits
20. ____ Louis de Borglie — b. explained black body radiation by using energy quanta
21. ____ Max Planck — c. developed the uncertainty principle
22. ____ Werner Heisenberg — d. proposed matter waves and calculated their wavelengths

Fill-in-the-Blank Questions (2 Points Each Answer)

23. To calculate speed, divide distance by ________________.

24. Momentum is the mass of an object times its ________________.

25. James Prescott Joule found how mechanical energy due to motion compares to ________________ energy.

26. The three factors that determine the heat contained in an object are type of substance, mass, and ________________.

27. The three properties of a sound are frequency, intensity, and ________________.

28. To reduce the heating effect of electricity in wires, the current is reduced but the ________________ is increased.

29. The Heisenberg uncertainty principle states that the precise position, mass, and ________________ cannot be determined exactly.

Multiple Choice Questions (2 Points Each Question)

30. Physics is the science that explores how energy acts on
    A. heat B. light C. matter D. sound

31. A feather and lump of lead will fall at the same speed in
    A. a high speed wind tunnel B. the atmosphere C. a vacuum D. water

32. Acceleration is found by dividing which of the following by the change in time?
    A. average velocity B. distance
    C. gravity D. change in speed)

33. Inertia is a property of matter that resists changing its
    A. electric charge B. mass C. momentum D. velocity

34. Heat is a type of A. energy B. force C. matter D. temperature

35. The two most common substances used in thermometers are colored alcohol and
    A. cooking oil B. ethylene glycol C. mercury D. molten salt

36. The scientist who discovered that pure water has a fixed boiling and freezing temperature was
    A. Anders Celsius B. Antoine Lavoisier
    C. Daniel Fahrenheit D. John Dalton

37. Density is equal to mass divided by
    A. area B. pressure C. volume D. weight

38. The frequency of a sound is known as its
    A. amplitude B. color C. pitch D. velocity

39. The study of sound is known as
    A. acoustics B. astronomy C. mechanics D. thermodynamics

40. The loudness of sound is measured in
    A. candles B. decibels C. joules D. watts

41. The force that pushes electrons around a circuit is

A. resistance B. charge C. current D. voltage

42. The scientist who developed four equations that summarized electromagnetism was
   A. Albert Einstein B. Isaac Newton
   C. James Clerk Maxwell D. Michael Faraday

43. The first scientist to generate electromagnetic waves was
   A. Arthur Compton B. Guglielmo Marconi
   C. Michael Faraday D. Rudolf Hertz

44. The first person to state that uranium could undergo fission and produce a self-sustaining chain reaction was
   A. Albert Einstein B. Enrico Fermi
   C. Franklin D. Roosevelt D. Lise Meiter

45. The purpose of a moderator is to
   A. absorb neutrons B. cool the reactor C. generate heat D. slow neutrons

Short Answer Questions (4 Points Each Question)

46. State the second law of motion:

47. State the third law of motion:

48. State Ohm's law:

True and False (2 Points Each)

49. T F Doubling mass of a moving object doubles its kinetic energy.

50. T F Doubling velocity of a moving object doubles its kinetic energy.

51. T F Scientists are unable to measure temperatures greater than 1,700°F.

52. T F Heat is the motion of atoms and molecules.

53. T F Thales of Melitus discovered that amber could be given a charge of static electricity.

54. T F An object with a positive charge has more protons than electrons.

Applied Learning Activities (1 Point Each Answer)

55. Label the four forces acting on an airplane in flight.

56. Write the numbers 1 to 4 in the blanks to rank these waves in order from lowest frequency (longest wavelength) to highest frequency (shortest wavelength):

_____ blue visible light

_____ AM radio waves

_____ X rays

_____ infrared light

	Exploring Biology	Quiz 1	Scope:	Total score:	Name
	Concepts & Comprehension		Chapters 1–3	____of 100	

Matching (3 Points Each Answer)

1. a. animal ____ non-living genetic material that only comes alive inside a living cell
   b. bacteria ____ multi-cellular life that can move and has sense organs
   c. fungi ____ multi-cellular life that includes mushrooms
   d. plant ____ multi-cellular life that makes food by photosynthesis
   e. protista ____ single-celled life that includes paramecium, amoeba, and euglena
   f. virus ____ single-celled life without a nucleus; one form causes Black Death (plague)

Fill-in-the-Blank Questions (4 Points Each Answer)

2. The single most deadly protozoa disease is ________________.
3. Lichen is a layer of algae sandwiched between two layers of ________________.
4. The Strait of Gibraltar opens from the Mediterranean Sea into the ________________ Ocean.
5. Another name for eyeteeth is ________________ teeth.

Multiple Choice Questions (4 Points Each)

6. To keep mushrooms in the plant kingdom, scientists described mushrooms as plants without
   A. cell walls B. chlorophyll C. seeds D. sunlight
7. Today, biologists classify mushrooms as members of the
   A. animal kingdom B. bacteria kingdom C. fungi kingdom D. plant kingdom
8. The above-ground stalk and umbrella of a mushroom is used to
   A. absorb carbon dioxide B. catch sunlight
   C. release spores D. sense the presence of enemies
9. What do yeast cells consume as food?
   A. alcohol B. carbon dioxide C. sugar D. vinegar
10. Fungi that are growing on bread and are just visible to the unaided eye and look like miniature mushrooms are most likely:
    A. lichen B. mold C. truffle D. yeast
11. The one that can change its shape is the
    A. amoeba B. euglena C. giardiasis D. paramecium
12. Single-celled algae that surround their cell wall with a coating of silicon dioxide are:
    A. anaerobic bacteria B. diatoms C. giardiasis cysts D. macrobiotic crust
13. The first group of sailors rumored to have discovered the Canary Islands were
    A. Greek B. Phoenician C. Roman D. Spanish

14. The Canary Islands were named for
A. birds B. cats C. dogs D. pigs

15. The English gold coin, the guinea, came from gold found in
A. Guinea along the west coast of Africa B. Morocco in Northern Africa
C. New Guinea, an island north of Australia D. the northern part of South America

16. The one who wrote Science of Botany was
A. a member of the Royal Society B. Aristotle
C. Carl Linnaeus D. Plato

Underline the Correct Answer (2 Points Each Answer)

17. Louis Pasteur realized yeast cells were alive when he saw them (A. cause milk to sour, B. grow and reproduce).

18. The first scientist to see the little life in a drop of canal water was (A. Robert Hooke, B. Anton van Leeuwenhoek).

19. At first, a euglena was called a plant because it (A. could not move, B. had chlorophyll).

20. Aristotle classified dolphins as (A. fish, B. mammals).

Applied Learning Activity (2 Points Each Answer)

The classification for mountain lion, *puma concolor* is: (select from animal, carnivore, chordata, concolor, feline, mammalia, puma)

21. kingdom:

22. phylum:

23. class:

24. order:

25. family:

26. genus:

27. species:

	Exploring Biology Concepts & Comprehension	Quiz 2	Scope: Chapters 4–7	Total score: ____of 100	Name

Fill-in-the-Blank Questions (2 Points Each Answer)

1. ______________________ were carried along the Silk Road from China to Spain and then to California.
2. The process by which plants can grow from a part of the parent plant such as a cutting from the stem or root is called ______________ reproduction.
3. Plants that take two years to produce seeds are known as ______________.
4. The white potato is what part of the plant? ______________
5. The reason cattle can digest grass is because they have ______________ in their digestive system.
6. The simple sugar ready for use by cells is ______________.
7. The three elements found in both carbohydrates and fats are ______________, ______________, and ______________ (any order).
8. To what state did Luther Burbank move after leaving New England? ______________
9. Prunes are partially dried fruit from a special type of ______________ tree.
10. In Carver's day, the important crops of the South were corn and ______________.

Multiple Choice Questions (3 Points Each)

11. Some plants have seeds with barbs, which are used for what purpose?
    A. as a way to be transported elsewhere B. to protect the growing sprout
    C. to provide food for the embryo D. to delay germination until spring
12. What plant has seeds that can be carried thousands of miles by ocean currents?
    A. coconut B. cottonwood C. dandelion D. tumbleweed
13. Starch is a type of (A. carbohydrate, B. fat, C. indigestible cellulose, D. protein).
14. The sugar found in mother's milk is (A. fructose B. glucose C. lactose D. maltose).
15. The one used for growth and repair of the body is
    A. carbohydrates B. fats C. glycogen D. proteins
16. Molar teeth are designed to (A. cut food B. grasp food C. mash food D. taste food
17. The stomach digests food by
    A. electrical impulses similar to a microwave B. chemically changing the food
    C. heating it and softening it D. mechanically mashing food
18. The chemical digestion of fats begins in the
    A. large intestine B. mouth C. small intestine D. stomach
19. The digested form of protein is
    A. amino acids B. enzymes C. fatty acids D. glucose

20. Taste buds are most sensitive to a (A. bitter taste B. mustard taste C. sour taste D. sweet taste

21. The four essential elements that plants need are carbon, oxygen, hydrogen, and:
A. iron B. nitrogen C. phosphorus D. sulfur

22. Carver suggested farmers plant peanuts and sweet potatoes because:
A. nitrogen-fixing bacteria grew along their roots
B. they had fine roots that prevented soil erosion
C. they produced far more income than cotton or corn
D. they released a chemical that killed harmful boll weevils

Underline the Correct Answer (3 Points Each Answer)

23. Most plants produce seeds in (A. the fall, B. early spring).

24. Spores are produced by plants that (A. flower, B. do not flower).

25. The one that is longer is the (A. large intestine, B. small intestine).

26. While in New England, Luther Burbank took sweet corn to market first because he (A. used a compost hotbed, B. imported them from California).

27. Most farmers plant potatoes by (A. planting potato seeds, B. planting a piece of a potato with an eye in it).

28. To speed up the development of prune trees, Luther Burbank first grew (A. almond trees, B. prune trees) in a greenhouse in a hotbed.

Short Answer (4 Points)

29. Why did George Washington Carver do laundry, ironing, and gardening rather than farm work?

Applied Learning Activity (18 Points Total: 2 Point Each Answer)

30–38. Anatomy of the stomach: identify body, cardia, circular muscle layer, duodenum, esophagus, longitudinal muscle layer, pyloric sphincter, oblique muscle layer, and rugae

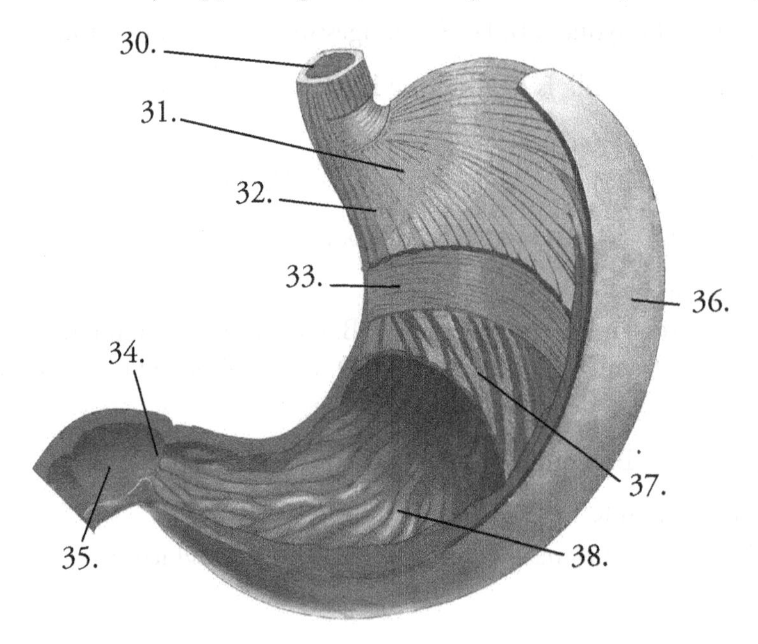

	Exploring Biology Concepts & Comprehension	Quiz 3	Scope: Chapters 8–11	Total score: ____of 100	Name

Matching (2 Points Each Answer)

1. Match the number of legs with the type of arthropod.

___ six	a. centipedes
___ eight	b. crabs and lobsters
___ ten	c. insects
___ one pair of legs per body segment	d. millipedes
___ two pair of legs per body segment	e. spiders and ticks

2. ___ The most common venomous snake found in the eastern United States

___ is cottony white inside its mouth

___ can weigh 20 pounds and is cooked as food

___ has glossy red, yellow, and black bands along its body

___ is the largest venomous snake

___ is a constrictor and is the largest snake by weight

a. anaconda b. copperhead c. coral snake
d. diamondback rattlesnake e. king cobra f . moccasin

Fill-in-the-Blank Questions (4 Points Each Answer)

3. All arthropods have a hard outer covering known as an ____________________.

4. Amphibians can breathe through gills, lungs, and ________________.

5. The lizard that can change color is the ________________.

Multiple Choice Questions (4 Points Each Question)

6. The spider that has a distinctive red hourglass pattern on the underside of the abdomen is the
A. female black widow B. male brown recluse C. orb weaver D. tarantula

7. Spider silk is stronger than silkworm silk because spider silk
A. contains a small strand of steel B. has a different composition
C. is consistent in thickness without weak spots D. is thicker and shorter

8. The one that is NOT a member of Class Arachnid is
A. grasshopper B. scorpion C. spider D. tick

9. The one that is a vertebrate is A. coral B. fish C. lancet D. snail

10. The classification of vertebrate is a A. class B. kingdom C. phylum D. subphylum

11. Amphibians include frogs, toads, and A. catfish B. goldfish C. salamanders D. salmon

Underline the Correct Answer (2 Points Each Answer)

12. The prefix arthro in arthropod means (A. foot, B. joint).

13. The one that is a type of insect is (A. cricket, B. shrimp).

14. The main goal of adult insects is to (A. eat as much food as possible to survive the winter, B. mate and reproduce).

15. A spider can be described as an arthropod with (A. six legs and three body segments, B. eight legs and two body segments).

16. A brown recluse is most likely to (A. hunt during the day, B. hunt at night).

17. The phrase that describes a parasite is (A. to eat at someone else's table, B. to live in two different places).

18. Fish are (A. cold blooded, B. warm blooded).

19. Amphibians are (A. cold blooded, B. warm blooded).

20. Biologists believe the number of amphibians is on the (A. rise, B. decline).

Short Answer (4 Points)

21. Why must arthropods molt?

22. What special sense organ makes it possible for a school of fish to turn together?

23. Why do snakes flick their forked tongues in and out?

Applied Learning Activity (12 Points Total: 2 Point Each Answer)

24–29. Snakehead anatomy: identify accessory gland, compressor muscle, fang, primary venom duct, secondary venom duct, and venom gland

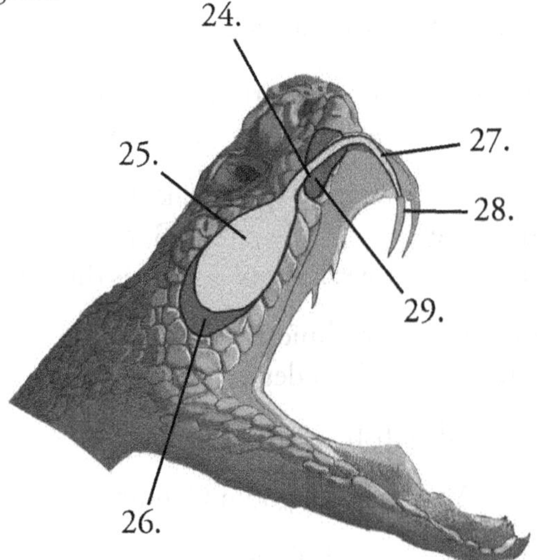

	Exploring Biology / Concepts & Comprehension	Quiz 4	Scope: Chapters 12–14	Total score: ____of 100	Name

Matching (2 Points Each Answer)

1. ___ Can mimic human speech
   ___ Used to hunt small game
   ___ Thick, heavyset, and flightless, this bird is extinct
   ___ Once numerous in the United States, now extinct
   ___ Has a long, thin beak that is like a hollow straw
   ___ Has a long, dagger-like beak used to spear fish
   ___ Migrates from Arctic to Antarctic
   ___ Large bird capable of killing a human

a. anhinga
b. Arctic tern
c. cassowary
d. dodo
e. falcon
f. hummingbird
g. parrot
h. passenger pigeon

Fill-in-the-Blank Questions (4 Points Each Answer)

2. Birds grown for human consumption are referred to as ______________.

3. Rather than teeth for crushing food, birds have a ______________ filled with grit and small stones.

4. The mammal that can fly is the ______________.

5. An example of a mammal that spends most of its time underground is the ______________.

Multiple Choice Questions (4 Points Each Question)

6. The one that is NOT true of mammals is that
   A. mammals are cold-blooded
   B. mammals are vertebrates
   C. mammals have a four-chambered heart
   D. mammals have sweat glands

7. The man who glued together parts to make a feathered dinosaur fossil did it:
   A. to embarrass his employer
   B. to gain entry into the United States from China
   C. to make the fossil more valuable
   D. to prove his skill as a fossil hunter

8. The Piltdown man deception began with the discovery of:
   A. broken pottery found in a trunk
   B. detailed cave paintings
   C. part of a skull
   D. the remains of a leg bone

9. The phony bones of Piltdown man were uncovered in:
   A. England B. France C. Germany D. Spain

10. The classification of humans, *Homo sapiens*, means:
    A. caveman B. dawn man C. upright man D. wise man

Underline the Correct Answer (2 Points Each Answer)

11. Birds are (A. cold-blooded, B. warm-blooded).

12. Compared to other vertebrates, birds have hearts that beat (A. less, B. more) quickly.

13. The milk that contains more fat is the milk of (A. horses, B. seals).

14. The word *nocturnal* means (A. active at night, B. hunt by sound echo).

15. Animals such as squirrels and beavers are examples of (A. canines, B. rodents).

16. Cats are mammals, in the order carnivore, and belong to the family (A. canine, B. feline).

17. Cheetahs catch their prey primarily by (A. pouncing on them with a sudden leap, (B. chasing them down).

18. A deliberate deception designed to gain money or something of value is a (A. fraud, B. hoax).

19. A paleontologist studies (A. past life, B. how animals migrate).

20. When Xu Xing, the Chinese scientist, found the other fossil slab in China, he discovered that it (A. did not match, B. was in far better shape) than the one in the United States.

True and False (2 Points Each)

21. T F The one feature that is true of all mammals is that the females provide milk for the young.

22. T F A four-chambered heart is actually two separate blood pumps.

23. T F The fact that the feathered dinosaur fossil was a fraud was discovered before it was made public.

24. T F Archaeologists study past human life.

25. T F Robert Virchow described the bones found in the Neander Valley as being from a short but stout human being.

26. T F The only tools ever found by the Neanderthals were stone clubs.

Short Answer (4 Points Each)

27. What birds did the British use to fly messages during the war with Napoleon?

28. Why do birds eat seeds and insects rather than grasses?

29. Why are the platypus and spiny anteater, who lay eggs rather than give live birth, classified as mammals?

30. Why did Maria see the painted bulls on the ceiling, but her father had overlooked them?

Matching (1 Point Each Answer)

1. ___ non-living genetic material that only comes alive inside a living cell
   ___ multi-cellular life that can move and has sense organs
   ___ multi-cellular life that includes mushrooms
   ___ multi-cellular life that makes food by photosynthesis
   ___ single-celled life that includes paramecium, amoeba, and euglena
   ___ single-celled life without a nucleus; one form causes Black Death (plague)

   a. animal
   b. bacteria
   c. fungi
   d. plant
   e. protista
   f. virus

2. Match the number of legs with the type of arthropod.
   ___ six
   ___ eight
   ___ ten
   ___ one pair of legs per body segment
   ___ two pair of legs per body segment

   a. centipedes
   b. crabs and lobsters
   c. insects
   d. millipedes
   e. spiders and ticks

3. ___ The most common venomous snake found in the eastern United States
   ___ is cottony white inside its mouth
   ___ can weigh 20 pounds and is cooked as food
   ___ has glossy red, yellow, and black bands along its body
   ___ is the largest venomous snake
   ___ is a constrictor and is the largest snake by weight

   a. anaconda
   b. copperhead
   c. coral snake
   d. diamondback rattlesnake
   e. king cobra
   f. moccasin

4. ___ Can mimic human speech
   ___ Used to hunt small game
   ___ Thick, heavyset, and flightless, this bird is extinct
   ___ Once numerous in the United States, now extinct
   ___ Has a long, thin beak that is like a hollow straw
   ___ Has a long, dagger-like beak used to spear fish
   ___ Migrates from Arctic to Antarctic
   ___ Large bird capable of killing a human

   a. anhinga
   b. Arctic tern
   c. cassowary
   d. dodo
   e. falcon
   f. hummingbird
   g. parrot
   h. passenger pigeon

Fill-in-the-Blank Questions (2 Points Each Answer)

5. The single most deadly protozoa disease is ________________.

6. Lichen is a layer of algae sandwiched between two layers of ________________.

7. The Strait of Gibraltar opens from the Mediterranean Sea into the ________________ Ocean.

8. Another name for eyeteeth is ________________ teeth.

9. ____ were carried along the Silk Road from China to Spain and then to California.

10. The process by which plants can grow from a part of the parent plant such as a cutting from the stem or root is called ________________ reproduction.

11. Plants that take two years to produce seeds are known as ________________.

12. The white potato is what part of the plant? ________________

13. The reason cattle can digest grass is because they have ________________ in their digestive system.

14. The simple sugar ready for use by cells is ________________.

15. The three elements found in both carbohydrates and fats are ________________, ________________, and ________________ (any order).

16. To what state did Luther Burbank move after leaving New England? ________________

17. All arthropods have a hard outer covering known as an ____________________.

18. Rather than teeth for crushing food, birds have a ________________ filled with grit and small stones.

19. The mammal that can fly is the ________________.

Multiple Choice Questions (2 Points Each Question)

20. The above-ground stalk and umbrella of a mushroom is used to
    A. absorb carbon dioxide
    B. catch sunlight
    C. release spores
    D. sense the presence of enemies

21. What do yeast cells consume as food?
    A. alcohol B. carbon dioxide C. sugar D. vinegar

22. Fungi that are growing on bread and are just visible to the unaided eye and look like miniature mushrooms are most likely
    A. lichen B. mold C. truffle D. yeast

23. The one that can change its shape is the
    A. amoeba B. euglena C. giardiasis D. paramecium

24. The digested form of protein is
    A. amino acids B. enzymes C. fatty acids D. glucose

25. Taste buds are most sensitive to a
    A. bitter taste B. mustard taste C. sour taste D. sweet taste

26. The four essential elements that plants need are carbon, oxygen, hydrogen, and
    A. iron B. nitrogen C. phosphorus D. sulfur

27. Carver suggested farmers plant peanuts and sweet potatoes because:
    A. nitrogen-fixing bacteria grew along their roots
    B. they had fine roots that prevented soil erosion
    C. they produced far more income than cotton or corn
    D. they released a chemical that killed harmful boll weevils

Short Answer (3 Points Each)

28. Why must arthropods molt?

29. Why do snakes flick their forked tongues in and out?

30. What birds did the British use to fly messages during the war with Napoleon?

31. Why do birds eat seeds and insects rather than grasses?

32. Why are the platypus and spiny anteater, who lay eggs rather than give live birth, classified as mammals?

33. Why did Maria see the painted bulls on the ceiling, but her father had overlooked them?

Applied Learning Activity (1 Points Each Answer)

The classification for mountain lion, *puma concolor* is: (select from animal, carnivore, chordata, concolor, feline, mammalia, puma)

34. kingdom:

35. phylum:

36. class:

37. order:

38. family:

39. genus:

40. species:

Exploring Chemistry / Concepts & Comprehension	Quiz 1	Scope: Chapters 1–4	Total score: ____of 100	Name

Fill-in-the-Blank Questions (4 Points Each Answer)

1. Ancient people hammered the soft, pure iron from ________________ into useful tools.
2. Cast iron, steel, and wrought iron differ only in the amount of________________they contain.
3. Gold, silver, and ________________ are known as the coinage metals.
4. The seven ancient metals are gold, silver, copper, iron, tin, lead, and ____________.
5. The single most important compound of sulfur is ________________ acid.
6. The element that charcoal, coal, graphite, and diamond have in common is ________________.
7. The Middle Ages are also known as the ________________ Ages.
8. State the chemical formula: _____ water _____ carbon dioxide _____hydrochloric acid

Multiple Choice Questions (3 Points Each)

9. Charcoal is
   A. a meteorite that fell from the heavens
   B. a type of coal found in the earth
   C. made of almost pure oxygen
   D. wood that has been heated without oxygen
10. Which of these forms of iron is the purest?
    A. cast iron B. charcoal C. steel D. wrought iron
11. Steel is quenched by
    A. burying it in the earth
    B. heating it in an oven for several days
    C. heating and thrusting it into cold water
    D. holding it overhead for lightning to strike
12. The first metal mentioned in both the Old and New Testaments is
    A. copper B. gold C. iron D. tin
13. Bronze and brass are both alloys that contain
    A. copper B. gold C. iron D. silver
14. Ancient people made musical instruments of
    A. copper alloy B. iron and mercury C. sulfur and carbon D. tin and lead
15. The Statue of Liberty has a skin of
    A. copper B. gold C. steel D. zinc
16. Another name for mercury is
    A. calliston B. cuprum C. plumbum D. quicksilver
17. Goodyear discovered vulcanized rubber when he
    A. added carbon to sulfur
    B. heated sulfur with raw rubber
    C. put rubber under intense pressure
    D. treated rubber with sulfuric acid
18. The Royal Society was formed to
    A. buy scientific equipment
    B. communicate new ideas rapidly
    C. help write better science textbooks
    D. teach the king about science

19. Cavendish released the gas hydrogen by exposing metals to
   A. ammonia
   B. hydrochloric acid
   C. carbon dioxide
   D. intense heat and pressure

20. The name hydrogen means
   A. colorless
   B. lacking odor
   C. lighter than air
   D. water generator

21. If a mixture of hydrogen and oxygen are exposed to a flame
   A. a violent explosion results
   B. electricity is generated
   C. the fire goes out
   D. the mixture becomes dry

22. The gas that Priestley released by heating a mercury compound was
   A. carbon dioxide
   B. hydrogen
   C. nitrogen
   D. oxygen

True/False **(2 Points Each)**

23. T F The only purpose of carbon in smelting iron from its ore is so it will burn and supply heat.

24. T F A 14-carat gold ring is pure gold.

25. T F The primary goal of alchemists was to make gold from cheap metals.

26. T F Hydrogen is an abundant element on earth.

Applied Learning Activity

Give the chemical symbol and place each one correctly on the Periodic Table of Elements **(1 Point Each)**

27. _____ hydrogen 28. ____ carbon 29. _____ nitrogen 30. _____ oxygen 31. _____ chlorine

Place the correct symbol for questions 27 – 31 on the chart below. **(1 Point Each)**

Group IA	IIA	IIIB	IVB	VB	VIB	VIIB	VIIIB	VIIIB	VIIIB	IB	IIB	IIIA	IVA	VA	VIA	VIIA	VIIIA
1 1.0079																	2 4.00260 He Helium
3 6.941 Li Lithium	4 9.01218 Be Beryllium											5 10.81 B Boron	6 12.011	7 14.0067	8 15.9994	9 18.998403 F Fluorine	10 20.179 Ne Neon
11 22.98977 Na Sodium	12 24.305 Mg Magnesium											13 26.98154 Al Aluminum	14 28.0855 Si Silicon	15 30.97376 P Phosphorus	16 32.06 S Sulfur	17 35.453	18 39.948 Ar Argon
19 39.0983 K Potassium	20 40.08 Ca Calcium	21 44.9559 Sc Scandium	22 47.90 Ti Titanium	23 50.9415 V Vanadium	24 51.996 Cr Chromium	25 54.9380 Mn Manganese	26 55.847 Fe Iron	27 58.9332 Co Cobalt	28 58.70 Ni Nickel	29 63.546 Cu Copper	30 65.38 Zn Zinc	31 69.72 Ga Gallium	32 72.59 Ge Germanium	33 74.9216 As Arsenic	34 78.96 Se Selenium	35 79.904 Br Bromine	36 83.80 Kr Krypton
37 85.4678 Rb Rubidium	38 87.62 Sr Strontium	39 88.9059 Y Yttrium	40 91.22 Zr Zirconium	41 92.9064 Nb Niobium	42 95.94 Mo Molybdenum	43 (98) Tc Technetium	44 101.07 Ru Ruthenium	45 102.9055 Rh Rhodium	46 106.4 Pd Palladium	47 107.868 Ag Silver	48 112.41 Cd Cadmium	49 114.82 In Indium	50 118.69 Sn Tin	51 121.75 Sb Antimony	52 127.60 Te Tellurium	53 126.9045 I Iodine	54 131.30 Xe Xenon
55 132.9054 Cs Cesium	56 137.33 Ba Barium	57 138.9055 La Lanthanum	72 178.49 Hf Hafnium	73 180.9479 Ta Tantalum	74 183.85 W Tungsten	75 186.207 Re Rhenium	76 190.2 Os Osmium	77 192.22 Ir Iridium	78 195.09 Pt Platinum	79 196.9665 Au Gold	80 200.59 Hg Mercury	81 204.37 Tl Thallium	82 207.2 Pb Lead	83 208.9804 Bi Bismuth	84 (209) Po Polonium	85 (210) At Astatine	86 (222) Rn Radon
87 (223) Fr Francium	88 226.0254 Ra Radium	89 227.0278 Ac Actinium	104 (261) Unq (Unniquadium)	105 (262) Unp (Unnilpentium)	106 (263) Sg (Seaborgium)	107 (262) Uns (Unnilseptium)	108 (265) Uno (Unniloctium)	109 (266) Une (Unnilennium)									

KEY: atomic number 5; atomic mass 10.81; oxidation number 3; symbol B; electron configuration $1s^2 2s^2 2p^1$; name Boron

58 140.12 Ce Cerium	59 140.9077 Pr Praseodymium	60 144.24 Nd Neodymium	61 (145) Pm Promethium	62 150.4 Sm Samarium	63 151.96 Eu Europium	64 157.25 Gd Gadolinium	65 158.9254 Tb Terbium	66 162.50 Dy Dysprosium	67 164.9304 Ho Holmium	68 167.26 Er Erbium	69 168.9342 Tm Thulium	70 173.04 Yb Ytterbium	71 174.967 Lu Lutetium
90 232.0381 Th Thorium	91 231.0359 Pa Protactinium	92 238.029 U Uranium	93 237.0482 Np Neptunium	94 (244) Pu Plutonium	95 (243) Am Americium	96 (247) Cm Curium	97 (247) Bk Berkelium	98 (251) Cf Californium	99 (252) Es Einsteinium	100 (257) Fm Fermium	101 (258) Md Mendelevium	102 (259) No Nobelium	103 (260) Lr Lawrencium

	Exploring Chemistry / Concepts & Comprehension	Quiz 2	Scope: Chapters 5–8	Total score: ____of 100	Name

Fill-in-the-Blank Questions (2 Points Each Answer)

1. An arc light generates light as electricity jumps across the gap between two ______________ electrodes.

2. The smallest objects that have the chemical properties of an element are ______________.

Multiple Choice Questions (4 Points Each)

3. Alessandro Volta built the first
   A. arc lamp
   B. device to make electricity by friction
   C. battery to produce electric current
   D. miner's safety lantern

4. Current electricity is due to the motion of
   A. electrons B. frog legs C. neutrons D. protons

5. The one who unwisely tasted and sniffed new chemicals was
   A. Benjamin Franklin
   B. Humphry Davy
   C. Michael Faraday
   D. Robert Wood

6. Davy discovered potassium by treating potash with
   A. a hot arc lamp
   B. heat from a large burning lens
   C. carbon in a blast furnace
   D. electricity from a strong battery

7. Dmitri Mendeleev was born in
   A. China B. Russia C. Spain D. United States

8. The first college that Dmitri Mendeleev attended was one
   A. to prepare farmers
   B. to teach philosophy
   C. for chemists
   D. to train teachers

9. Mendeleev believed an organized table of the elements would
   A. discourage his students
   B. help his students
   C. show chemistry to be a difficult subject
   D. make him a lot of money

10. Dmitri Mendeleev set about organizing the table of elements by
   A. asking students to vote on their most popular element
   B. collecting samples of each element
   C. entering information about the elements in a computer
   D. writing information about the elements on note cards

11. To silence critics of his table, Mendeleev
   A. appealed to the government to arrest them
   B. asked his friends to argue his case
   C. left the country
   D. predicted the properties of three missing elements

12. The three most important tools for making advances in chemistry were electricity, the periodic law, and the A. law of buoyancy B. microscope C. spectroscope D. telescope

13. The first element discovered by the spectroscope was
A. cesium B. helium C. kryptonite D. uranium

14. Henry Cavendish combined oxygen with nitrogen by using
A. an electric spark B. high heat C. intense cold D. great pressure

15. The name helium means from
A. Helena, Montana B. the hills C. the sun D. uranium

16. The noble gas family is also known as the
A. empire of the sun family B. inert gases
C. Ramsay and Rayleigh families D. strange family

17. William Crookes did scientific experimentation
A. because he enjoyed it B. to discover a way to make synthetic gold
C. to earn money for his poor family D. to prove the worth of English research

18. A radiometer
A. compares the colors of different elements B. measures the strength of radiant energy
C. was an early form of the barometer D. was an early form of the radio

19. Most alpha particles entered the gold foil and
A. bounced out in all directions B. caused the gold to become radioactive
C. passed straight through D. were absorbed, never to be seen again

20. About 1 alpha particle in 10,000
A. bounced out in the same direction it entered B. caused a spark of light
C. passed through the gold foil D. turned into gold

True and False (3 Points Each)

21. T F Static electricity was unknown to the ancient Greeks.

22. T F Dmitri Mendeleev organized his table of the elements by atomic weight and valence.

23 T F Only one of Mendeleev's missing elements has been found.

24. T F Fraunhofer saw the lines in the spectrum while testing the quality of a prism.

25. T F Robert Bunsen invented the Bunsen burner.

26. T F Argon helps make light bulbs last longer.

27. T F According to Dalton, atoms of the same element are identical in all their properties, including their weight.

28. T F For many years chemists ignored John Dalton's atomic theory of matter and refused to accept it.

Exploring Chemistry / Concepts & Comprehension	Quiz 3	Scope: Chapters 9–12	Total score: ____of 100	Name

Fill-in-the-Blank Questions (4 Points Each Answer)

1. Timbuktu is located at the edge of the ________________ Desert.
2. The purple dye containing bromine was prepared by Phoenicians who lived in the city of ________________.
3. Carbon needs ________________ more electrons to have its electrons in the stable arrangement like the electrons of neon.
4. The number of chlorine atoms in carbon tetrachloride is ________________.
5. The elements found in organic compounds include hydrogen, oxygen, nitrogen, and ________________.

Underline the Correct Answer (2 Points Each Answer)

6. Chlorinating drinking water was first tried in London in 1897 to (A. prevent tooth decay, B. stop an outbreak of typhoid fever).
7. Sodium and chlorine families form compounds by (A. sharing electrons, B. electrical attraction).
8. Oxygen and hydrogen form the water molecule by (A. exchanging, B. sharing) electrons.
9. Liquid water (A. expands, B. contracts) when it changes into ice.
10. Ice (A. floats, B. sinks) in water.
11. Snow conducts heat (A. well, B. poorly).
12. In cool water, molecules move more (A. quickly, B. slowly) than in warm water.
13. The one that boils at a hotter temperature is (A. methane, B. water).
14. When burned with oxygen, natural gas releases carbon dioxide and (A. sulfuric acid, B. water).
15. Paraffin is an example of a (A. short, B. long) hydrocarbon chain.
16. Teflon is (A. sticky, B. slick).
17. Ethyl alcohol and dimethyl ether are examples of (A. polymers; B. isomers).
18. Salt is an example of a compound that is (A. formed by living organisms, B. found in the non-living environment).
19. Sugar is an example of a compound that is (A. formed by living organisms, B. found in the non-living environment).

Multiple Choice Questions (3 Points Each Question)

20. The plus sign, +, for the charged sodium atom, Na+, shows that it

    A. carries a negative charge B. has more protons than electrons

    C. is made of a single proton D. is too big to take part in chemical reaction

21. The charged sodium atom, Na+, and the charged chlorine atom, Cl-, will
A. attract one another B. come together and destroy one another
C. produce an electric current D. repel one another

22. Another name for the chlorine family is
A. acid makers B. alkali makers C. radioactive family D. salt makers

23. A diagram of a water molecule shows it as
A. a constantly shifting molecule that is never the same way twice B. a long chain
C. an oxygen atom face with hydrogen atom ears D. a molecule shaped like a pyramid

24. The Nile River continues to flow through deserts because
A. it is fed by desert springs B. it is fed by melting snow in the mountains
C. rain falls year-round in Egypt D. water is too heavy to evaporate

25. Water boils at
A. 0°C [32°F] B. 100°C [212°F]
C. -161°C [-258°F] D. various temperatures depending on the phase of the moon

26. Methane is also known as what kind of gas? A. marsh B. mustard C. poison D. tear

27. A hydrocarbon is a compound that contains only carbon and
A. chlorine B. fluorine C. hydrogen D. oxygen

28. Leo Baekeland discovered the substance he called Bakelite while trying to make a substitute for
A. a dye B. a perfume C. eye shadow D. shellac

29. When Leo Baekeland mixed carbolic acid and formaldehyde, the result was
A. a substance that clogged test tubesB. a substance with the smell of new-mown hay
C. a thin liquid D. an explosive gas

True and False (2 Points Each)

30. T F Although salt flavors food, it has no other purpose in the diet.

31. T F Water is the most common liquid on earth.

32. T F Wilson Bentley became wealthy from his hobby of photographing snowflakes.

33. T F The ozone layer reduces the effect of harmful ultraviolet rays from the sun.

34. T F Freon is used for refrigeration.

35. T F Whenever a liquid evaporates, it warms its surroundings.

36. T F Methanol is burned as a fuel in racing engines.

37. T F Chemical reactions follow a different set of rules in living things than they do in the laboratory.

38. T F No one has ever succeeded in making organic compounds in the laboratory.

39. T F After he became successful, William Henry Perkin retired from chemical research.

40. T F Although Bakelite was the first plastic, it was immediately replaced by better ones and proved a disappointing failure to Leo Baekeland.

Matching (2 Points Each)

1. ____ helped found the Royal Society. He also defined an element.
2. ____ was an eccentric English chemist who discovered hydrogen.
3. ____ was a Frenchman who burned a diamond. He stated the law of conservation of matter.
4. ____ stated the atomic theory of matter.
5. ____ used electricity to free sodium and other elements from their ores.
6. ____ suggested the use of chemical symbols for elements.
7. ____ was Davy's assistant who discovered benzene.
8. ____ was a French chemist who became a medical researcher.
9. ____made the first periodic table of the elements.
10. ____ discovered the family of inert gases.

a. Jöns Jakob Berzelius
b. Robert Boyle
c. Henry Cavendish
d. John Dalton
e. Humphry Davy
f. Michael Faraday
g. Antoine Laurent Lavoisier
h. Dmitri Ivanovich Mendeleev
i. Louis Pasteur
j. William Ramsay

Fill-in-the-Blank Questions (4 Points Each Answer)

11. The essential element in explosives is ________________.
12. Nitrocellulose is a combination of a ________________ compound with cellulose.

Multiple Choice Questions (4 Points Each Question)

13. Cellulose gives plant cells their
    A. ability to do photosynthesis
    B. color
    C. daily supply of water
    D. strong cell walls
14. Abraham Lincoln's great construction project was to connect California to the East Coast with
    A. an interstate highway
    B. a pony express route
    C. a railroad
    D. a telegraph line.
15. Diatoms have cell walls of
    A. cellulose
    B. dynamite
    C. nitroglycerin
    D. silica
16. Blasting caps are used
    A. for safe detonation of explosives
    B. for small explosions
    C. to contain an explosion
    D. to control the direction of an explosion

17. Silicon carbide was discovered while trying to make synthetic
A. diamond B. glass C. rubies D. Silly Putty

18. Stained glass was colored
A. because church officials did not want people to look outside during church services
B. to hide the fact that glass makers could not make clear glass
C. to prevent it from melting
D. because people did not like clear glass

19. A gem's value is due to its
A. beauty B. cost C. rarity D. all of the above

20. The Star of India is
A. a diamond B. a sapphire C. a synthetic ruby D. an opal

21. Silicon is an electric
A. conductor B. insulator C. semiconductor D. none of the above

22. France's Emperor Napoleon III had table sets made of
A. aluminum B. copper C. frozen nitrogen D. uranium

23. The six-pound metallic top to the Washington Monument is made of
A. aluminum B. gold C. silver D. uranium

24. The foil used to wrap foods is made of
A. aluminum B. steel C. copper D. zinc

25. Uranium ore is
A. galena B. hematite C. pitchblende D. sulfur dioxide

26. The radioactive element used in smoke detectors is
A. americium B. curium C. radium D. uranium

Underline the Correct Answer (2 Points Each Answer)

27. The powder in the expression "keep your powder dry" was (A. diatomaceous earth, B. gunpowder).

28. Guncotton would be described as being (A. a safe and effective explosive, B. unpredictable and capable of exploding without warning).

29. Ascanio Sobrero's reaction to nitroglycerin was to (A. announce its discovery at a chemical congress, B. keep it secret).

30. Silicon is the (A. most abundant, B. second most abundant) element in the earth's crust.

31. The one that can resist sudden changes in temperature is (A. glass, B. quartz).

32. The one better able to lubricate across extreme temperatures is (A. hydrocarbon motor oil, B. silicone oil).

33. The element found in computer chips is (A. carbon, B. silicon).

34. The one discovered first was the (A. planet Uranus, B. element uranium).

	Exploring Chemistry / Concepts & Comprehension	Test 1	Scope: Chapters 1–16	Total score: ____of 100	Name

Matching (1 Point Each)

1. ____ helped found the Royal Society. He also defined an element.
2. ____ was an eccentric English chemist who discovered hydrogen.
3. ____ was a Frenchman who burned a diamond. He stated the law of conservation of matter.
4. ____ stated the atomic theory of matter.
5. ____ used electricity to free sodium and other elements from their ores.
6. ____ suggested the use of chemical symbols for elements.
7. ____ was Davy's assistant who discovered benzene.
8. ____ was a French chemist who became a medical researcher.
9. ____made the first periodic table of the elements.
10. ____ discovered the family of inert gases.

a. Jöns Jakob Berzelius
b. Robert Boyle
c. Henry Cavendish
d. John Dalton
e. Humphry Davy
f. Michael Faraday
g. Antoine Laurent Lavoisier
h. Dmitri Ivanovich Mendeleev
i. Louis Pasteur
j. William Ramsay

Fill-in-the-Blank Questions (1 Point Each Answer)

11. Ancient people hammered the soft pure iron from ________________ into useful tools.
12. Cast iron, steel, and wrought iron differ only in the amount of ________________ they contain.
13. Gold, silver, and ________________ are known as the coinage metals.
14. The seven ancient metals are gold, silver, copper, iron, tin, lead, and ____________.
15. The single most important compound of sulfur is ________________ acid.
16. The element that charcoal, coal, graphite, and diamond have in common is ________________.
17. The Middle Ages are also known as the ________________ Ages.
18. State the chemical formula: _____ water _____ carbon dioxide _____ hydrochloric acid
19. An arc light generates light as electricity jumps across the gap between two ________________ electrodes.
20. The smallest objects that have the chemical properties of an element are ________________.
21. Timbuktu is located at the edge of the ________________ Desert.

22. The purple dye containing bromine was prepared by Phoenicians who lived in the city of ________________.

23. Carbon needs ________________ more electrons to have its electrons in the stable arrangement like the electrons of neon.

24. The number of chlorine atoms in carbon tetrachloride is ________________.

25. The elements found in organic compounds include hydrogen, oxygen, nitrogen, and ______________.

26. The essential element in explosives is ________________.

27. Nitrocellulose is a combination of a ________________ compound with cellulose.

Multiple Choice Questions (2 Points Each Question)

28. The Statue of Liberty has a skin of A. copper B. gold C. steel D. zinc

29. Another name for mercury is A. calliston B. cuprum C. plumbum D. quicksilver

30. Goodyear discovered vulcanized rubber when he
    A. added carbon to sulfur
    B. heated sulfur with raw rubber
    C. put rubber under intense pressure
    D. treated rubber with sulfuric acid

31. The Royal Society was formed to
    A. buy scientific equipment
    B. communicate new ideas rapidly
    C. help write better science textbooks
    D. teach the king about science

32. To silence critics of his table, Mendeleev
    A. appealed to the government to arrest them
    B. asked his friends to argue his case
    C. left the country
    D. predicted the properties of three missing elements

33. The three most important tools for making advances in chemistry were electricity, the periodic law, and the A. law of buoyancy B. microscope C. spectroscope D. telescope

34. The first element discovered by the spectroscope was
    A. cesium B. helium C. kryptonite D. uranium

35. Henry Cavendish combined oxygen with nitrogen by using
    A. an electric spark B. high heat C. intense cold D. great pressure

36. The charged sodium atom, Na+, and the charged chlorine atom, Cl-, will
    A. attract one another
    B. come together and destroy one another
    C. produce an electric current
    D. repel one another

37. Another name for the chlorine family is
    A. acid makers B. alkali makers C. radioactive family D. salt makers

38. A diagram of a water molecule shows it as
    A. a constantly shifting molecule that is never the same way twice
    B. a long chain
    C. an oxygen atom face with hydrogen atom ears
    D. a molecule shaped like a pyramid

39. The Nile River continues to flow through deserts because
A. it is fed by desert springs
B. it is fed by melting snow in the mountains
C. rain falls year-round in Egypt
D. water is too heavy to evaporate

40. A gem's value is due to its
A. beauty B. cost C. rarity D. all of the above

41. The Star of India is
A. a diamond B. a sapphire C. a synthetic ruby D. an opal

42. Silicon is an electric
A. conductor B. insulator C. semiconductor D. none of the above

43. France's Emperor Napoleon III had table sets made of
A. aluminum B. copper C. frozen nitrogen D. uranium

Underline the Correct Answer (1 Point Each Answer)

44. Chlorinating drinking water was first tried in London in 1897 to (A. prevent tooth decay, B. stop an outbreak of typhoid fever).

45. Sodium and chlorine families form compounds by (A. sharing electrons, B. electrical attraction).

46. Oxygen and hydrogen form the water molecule by (A. exchanging, B. sharing) electrons.

47. Silicon is the (A. most abundant, B. second most abundant) element in the earth's crust.

48. The element found in computer chips is (A. carbon, B. silicon).

True and False (1 Point Each Answer)

49. T F The only purpose of carbon in smelting iron from its ore is so it will burn and supply heat.

50. T F A 14-carat gold ring is pure gold.

51. T F The primary goal of alchemists was to make gold from cheap metals.

52. T F Hydrogen is an abundant element on earth.

53. T F Static electricity was unknown to the ancient Greeks.

54. T F Dmitri Mendeleev organized his table of the elements by atomic weight and valence.

55. T F Only one of Mendeleev's missing elements has been found.

56. T F Fraunhofer saw the lines in the spectrum while testing the quality of a prism.

57. T F Robert Bunsen invented the Bunsen burner.

58. T F Argon helps make light bulbs last longer.

59. T F According to Dalton, atoms of the same element are identical in all their properties, including their weight.

60. T F For many years chemists ignored John Dalton's atomic theory of matter and refused to accept it.

61. T F Although salt flavors food, it has no other purpose in the diet.

62. T F Water is the most common liquid on earth.

63. T F Wilson Bentley became wealthy from his hobby of photographing snowflakes.

64. T F The ozone layer reduces the effect of harmful ultraviolet rays from the sun.

65. T F Freon is used for refrigeration.

66. T F Whenever a liquid evaporates, it warms its surroundings.

67. T F Methanol is burned as a fuel in racing engines.

68. T F Chemical reactions follow a different set of rules in living things than they do in the laboratory.

69. T F No one has ever succeeded in making organic compounds in the laboratory.

70. T F After he became successful, William Henry Perkin retired from chemical research.

71. T F Although Bakelite was the first plastic, it was immediately replaced by better ones and proved a disappointing failure to Leo Baekeland.

Applied Learning Activity

72. Give the chemical symbol and place each one correctly on the Periodic Table of Elements. **(1 Point Each)**

_____ hydrogen _____ carbon _____ nitrogen _____ oxygen _____ chlorine

Place the correct symbol for question 72 on the chart below. **(1 Point Each)**

Group IA	IIA	IIIB	IVB	VB	VIB	VIIB	VIIIB			IB	IIB	IIIA	IVA	VA	VIA	VIIA	VIIIA
1 1.0079																	2 4.00260 He Helium
3 6.941 Li Lithium	4 9.01218 Be Beryllium											5 10.81 B Boron	6 12.011	7 14.0067	8 15.9994	9 18.998403 F Fluorine	10 20.1 Ne Neon
11 22.98977 Na Sodium	12 24.305 Mg Magnesium											13 26.98154 Al Aluminum	14 28.0855 Si Silicon	15 30.97376 P Phosphorus	16 32.06 S Sulfur	17 35.453	18 39.948 Ar Argon
19 39.0983 K Potassium	20 40.08 Ca Calcium	21 44.9559 Sc Scandium	22 47.90 Ti Titanium	23 50.9415 V Vanadium	24 51.996 Cr Chromium	25 54.9380 Mn Manganese	26 55.847 Fe Iron	27 58.9332 Co Cobalt	28 58.70 Ni Nickel	29 63.546 Cu Copper	30 65.38 Zn Zinc	31 69.72 Ga Gallium	32 72.59 Ge Germanium	33 74.9216 As Arsenic	34 78.96 Se Selenium	35 79.904 Br Bromine	36 83.80 Kr Krypton
37 85.4678 Rb Rubidium	38 87.62 Sr Strontium	39 88.9059 Y Yttrium	40 91.22 Zr Zirconium	41 92.9064 Nb Niobium	42 95.94 Mo Molybdenum	43 (98) Tc Technetium	44 101.07 Ru Ruthenium	45 102.9055 Rh Rhodium	46 106.4 Pd Palladium	47 107.868 Ag Silver	48 112.41 Cd Cadmium	49 114.82 In Indium	50 118.69 Sn Tin	51 121.75 Sb Antimony	52 127.60 Te Tellurium	53 126.9045 I Iodine	54 131.30 Xe Xenon
55 132.9054 Cs Cesium	56 137.33 Ba Barium	57 138.9055 La Lanthanum	72 178.49 Hf Hafnium	73 180.9479 Ta Tantalum	74 183.85 W Tungsten	75 186.207 Re Rhenium	76 190.2 Os Osmium	77 192.22 Ir Iridium	78 195.09 Pt Platinum	79 196.9665 Au Gold	80 200.59 Hg Mercury	81 204.37 Tl Thallium	82 207.2 Pb Lead	83 208.9804 Bi Bismuth	84 (209) Po Polonium	85 (210) At Astatine	86 (222) Rn Radon
87 (223) Fr Francium	88 226.0254 Ra Radium	89 227.0278 Ac Actinium	104 (261) Unq (Unnilquadium)	105 (262) Unp (Unnilpentium)	106 (263) Sg (Seaborgium)	107 (262) Uns (Unnilseptium)	108 (265) Uno (Unniloctium)	109 (266) Une (Unnilennium)									

KEY: atomic number 5; atomic mass 10.81; oxidation number 3; symbol B; electron configuration $1s^2 2s^2 2p^1$; name Boron

58 140.12 Ce Cerium	59 140.9077 Pr Praseodymium	60 144.24 Nd Neodymium	61 (145) Pm Promethium	62 150.4 Sm Samarium	63 151.96 Eu Europium	64 157.25 Gd Gadolinium	65 158.9254 Tb Terbium	66 162.50 Dy Dysprosium	67 164.9304 Ho Holmium	68 167.26 Er Erbium	69 168.9342 Tm Thulium	70 173.04 Yb Ytterbium	71 174.967 Lu Lutetium
90 232.0381 Th Thorium	91 231.0359 Pa Protactinium	92 238.029 U Uranium	93 237.0482 Np Neptunium	94 (244) Pu Plutonium	95 (243) Am Americium	96 (247) Cm Curium	97 (247) Bk Berkelium	98 (251) Cf Californium	99 (252) Es Einsteinium	100 (257) Fm Fermium	101 (258) Md Mendelevium	102 (259) No Nobelium	103 (260) Lr Lawrencium

Answer Keys

Chapter 1

1. F, 2. F
3. So the calendar will match the seasons. Or, so the calendar year will be the same length as the solar year.

4. a, 5. c, 6. e, 7. d, 8. a, 9. b

10. 969 years x 365 days per year = 353,685 days

11. 120 days, 72 days, 18 days, 6 days (divide 360 by 3, 5, 20, and 60)

12. divide the population by 1,461

Chapter 2

1. A, 2. F, 3. D, 4. A, 5. F,
6. C, 7. D, 8. B, 9. B, 10. B

11. 6:30 a.m. The second watch began at 4:00 a.m. Each bell is ½ hour. Five bells are 2½ hours: 4:00 a.m. + 2 hr. 30 min. = 6:30 a.m.

12. Answer varies depending on actual heart rate. For 72 beats per min: 72 beats per min. x 60 min. per hr. x 24 hr. per day = 103,680 beats per day

13. 8 hours. One way to solve the problem is to change to military time and subtract — 9:00 a.m. is 0900 and 5:00 p.m. is 1700: 1700 – 0900 = 0800 or 8 hours.

14. one hour later, 4:00 p.m. MST is 5:00 p.m. CST

Chapter 3

1. B, 2. A, 3. C, 4. A, 5. F
6. B, 7. 5,280, 8. pound, 9. B
10. A, 11. A, 12. B, 13. A

14. 60 inches, 5 feet. Multiplying 15 hands by 4 inches per hand gives 60 inches. Sixty inches is equal to five feet: 60 in ÷ 12 in. per ft. = 5 ft.

15. Answer varies. Multiply weight in pounds by the conversion factor of 16 ounces per pound.

16. 5.499 miles or about 5.5 miles. Divide 29,035 feet by the conversion factor of 5,280 feet per mile.

Chapter 4

1. B, 2. B, 3. F, 4. D, 5. T
6. B, 7. B, 8. B, 9. A, 10. D
11. D, 12. F, 13. C

Chapter 5

1. B, 2. T, 3. F, 4. B, 5. D
6. F, 7. A, 8. D, 9. A, 10. A
11. e, 12. a, 13. d , 14. c, 15. b
16. d, 17. e, 18. f, 19. b, 20. a
21. c

22. 140 tiles. The area of the room is 140 square feet, A = L x W = 14 ft. x 10 ft. = 140 sq. ft., and each tile covers one square foot, so 140 tiles are needed.

Chapter 6

1. B
2. squares, square
3. A, 4. D, 5. A, 6. e, 7. d, 8. c
9. a, 10. b, 11. b, 12. d, 13. a
14. c

Chapter 7

1. T, 2. F, 3. T, 4. T, 5. F, 6. T
7. T, 8. F, 9. F, 10. F, 11. F
12. F, 13. F

Chapter 8

1. B, 2. C, 3. A, 4. A, 5. A
6. C, 7. B, 8. A, 9. B, 10. F
11. T, 12. B, 13. C
14. 233 = 89 + 144
15. F, 16. b, 17. c, 18. a, 19. d
20. e

Chapter 9

1. B, 2. B, 3. T, 4. T, 5. F, 6. T
7. D, 8. C, 9. B, 10. A, 11. A
12. F, 13. B, 14. D, 15. a, 16. g
17. b, 18. e, 19. c, 20. f, 21. d

Chapter 10

1. T, 2. C, 3. A, 4. B, 5. T, 6. F

7. C, 8. B, 9. C, 10. b, 11. d

12. c, 13. a, 14. b, 15. d, 16. a

17. c

Chapter 11

1. A, 2. A, 3. C, 4. B, 5. B

6. C, 7. B, 8. F, 9. e, 10. b

11. a, 12. d, 13. c

14. 17,576,000 — Any one of 26 letters (A through Z) can be chosen to fill the first three positions. Any one of 10 digits (zero through nine) can be chosen to fill the second group of three positions: 26 x 26 x 26 x 10 x 10 x 10 = 17,576,000.

Chapter 12

1. B, 2. F, 3. T, 4. F, 5. T

6. three, ten, 0.477

7. D, 8. B, 9. T, 10. B

11. central processing unit

12. A, 13. h, 14. c, 15. d, 16. f

17. g, 18. a, 19. e, 20. b

Chapter 13

1. 9, 2. F, 3. A, 4. B, 5. C, 6. T

7. B, 8. D, 9. F, 10. F, 11. D

12. B, 13. B, 14. A, 15. C, 16. T

17. A

18. About 3.8 seconds: 26,747 characters / 7,000 bytes (characters) per second = 3.821 seconds

Chapter 14

Puzzle 1: Multiplying by Seven. You discover that 7 x 142,857 = 999,999. The reason that the answer suddenly goes to all nines is surprising, but only because the source of the number 142,857 was not given. The number 100,000 divided by 7 is a repeating decimal: 1,000,000 ÷ 7 = 142,857.142857142857 . . . with the pattern 142857 repeating. Now 1,000,000 ÷ 7 x 7 = 1,000,000, so 7 x 142,857.142857142857 . . . would also be 1,000,000. However, if only the first group is used, then 7 x 142,857 = 999,999. The final one needed to roll the number up to one million is missing because the repeating part of the decimal fraction was not used.

Puzzle 2: Multiplying by 99. The left-most digit (the one in the 100s place) in the answer goes from 1 to 8 while the right-most digit (the one in the 1s place) goes from 8 to 1. This result is easy to see once you think that the number 99 is 100 – 1. Multiplying by two gives 200 – 2 = 198; by three gives 300 – 3 = 297; by four gives 400 – 4 = 396 and so on.

Puzzle 3: On the Road to St. Ives. Only one person is on the road to St. Ives. Instead of a large number, this is a trick question. The last line of the rhyme asks how many were going to St. Ives. The wives and the things they carried were going away from St. Ives. The narrator (the "I" person) is the one going to St. Ives.

Puzzle 4: Send More Money.

```
  S E N D
+ M O R E
M O N E Y

  9 5 6 7
+ 1 0 8 5
1 0 6 5 2
```

M in MONEY must be 1 because even with a carry, the sum of S and M is less than 20. O must be zero because M is 1 and S must be 9 or 8 with a carry of 1. Either value for S forces O to be zero. Now that we know O is zero, S must be 9 because 8 is too small even with a carry of 1 to be ten.

The total of N and R must be greater than 10 to give a carry of 1; otherwise E plus zero and no carry would equal R, and two letters cannot have the same value. N is one more than E. Trying the remaining numbers in E + 0 = N and N + R = 10 + E shows that R is 8, N is 6 and E is 5. D = 7, E = 5, M = 1, N = 6, O =

0, R = 8, S = 9, Y = 2.

Puzzle 5: For More Study — The 3N + 1 Problem. The sequence is 18, 9, 28, 14, 7, 22, 11, 34, 17, 52, 26, 13, 40, 20, 10, 5, 16, 8, 4, 2, 1.

Puzzle 6: Samson's Riddle. Judges 14:14 tells how Samson came to think of the puzzle. He found a lion's carcass with a honeycomb from a beehive inside. Out of the eater (lion), something to eat, out of the strong, something sweet (honey).

Puzzle 7: Sock Puzzle. One extra sock is enough. If the socks he is wearing match, he does not need the spare one. If his socks are not alike, then one will be the same color as the one he is carrying, giving him a matching pair.

Puzzle 8: River Crossing. First, take the goat across and leave him on the far bank. The carrots will be safe left alone with the wolf. Return for the carrots and leave them on the far bank but pick up the goat and return with him to the near bank. Leave the goat and paddle the wolf across. Leave the wolf with the carrots and return for the goat.

Puzzle 9: Dürer's Number Square. To get started, you should figure out the sum for each row, column, and diagonal. The numbers 1 through 9 sum to 45, and each row (and column and diagonal) must sum to 15 (45/3 = 15). One possible arrangement is:

8	1	6
3	5	7
4	9	2

The other seven solutions are merely rotations and mirror images of this solution.

Puzzle 10: Grass to Milk. BESSIE is the cow's name.

Multiply prime numbers: 7 x 17 x 23 = 2,737.

Find area: A = L x W = 201 ft. x 201 ft. = 40,401 sq. ft.

Seconds in a day: 60 sec./min. x 60 min./hr. x 24 hr./day = 86,400 sec.

Speed of light 186,000 mi./sec.

Adding the numbers (ignore the units): 2,737 + 40,401 + 86,400 + 186,000 = 315538.

Exploring the World of Physics — Worksheet Answer Keys

Chapter 1

1. C. matter
2. F. They seldom did experiments.
3. T
4. F. length and weight
5. C. a vacuum
6. time
7. D. rolled them down a ramp
8. D. change in speed
9. A. 32 ft./sec^2
10. 53 miles per day. Average speed is the distance divided by the time: speed = distance/time = 3,710 miles/70 days = 53 mi./da.
11. 7.5 mi./hr. × sec. Acceleration = (change in speed)/(change in time) = (60 mi/hr) (8 sec) = 7.5 mi./hr. × sec.
12. 31.8 ft./sec. $v_f = a \times t$ = (5.3 ft./sec^2) (6 sec.) = 31.8 ft./sec.

Chapter 2

1. F. velocity includes direction as well as speed

2. T
3. F. force must act on a moving object to slow it to a stop
4. A. friction
5. B. Galileo
6. D. velocity
7. F. every object has inertia
8. T
9. The acceleration of an object is directly proportional to the force acting on it and inversely proportional to its mass.
10. To every action there is an equal and opposite reaction.
11. velocity
12. T, 13. e, 14. a, 15. c, 16. b
17. d, 18. f

Chapter 3

1. T
2. B — elliptical
3. A — faster
4. The straight line joining a planet with the sun sweeps out equal areas in equal intervals of time.
5. T
6. B — Kepler
7. F — his father had died and his mother was poor
8. T
9. A — 3,600 times weaker ($60^2 = 3,600$)
10. F. to all objects
11. B
12. product, square
13. F. Scientists believe they have found about 100 planets around distant stars.

Chapter 4

1. force
2. A. Archimedes
3. A. effort
4. B. a lever
5. F. at either end or anywhere in between
6. B. reduced
7. M.A. = 9. Mechanical advantage = ramp length/ramp height = run rise = 9 miles/1 mile = 9
8. C. a wheel and axle
9. T
10. F — it is less
11. B. 18 wheeler truck
12. B. 100

Chapter 5

1. F — only since the 1800s
2. B — energy
3. T
4. D — work
5. energy
6. D — work
7. heat
8. F — The desk must move for work to be done.
9. B — power
10. C — horsepower and watt
11. A — kinetic
12. T
13. F — it becomes four times as great
14. B — doubling its velocity
15. B — potential energy
16. A — heat

Chapter 6

1. A — energy
2. temperature
3. B — water
4. B — expand
5. C — mercury
6. F — They can measure higher temperatures with electrical conductivity and color of light emitted by glowing substance.
7. C — Daniel Fahrenheit
8. A — higher
9. T
10. A — kinetic
11. A — conduction

12. A — copper
13. A — poorly
14. B — convection
15. T
16. T
17. B — greatly different
18. T
19. 0.058 or about six percent
Efficiency = $(T_1 - T_2)/T_1$ = (291 K – 274 K)/(291 K) = 17/291 = 0.058 or less than six percent

Chapter 7

1. F — Because it will snap back to its original shape.
2. T
3. force
4. B — reduces
5. A — height
6. T
7. C — volume
8. B — decrease
9. B — square root
10. b, 11. d, 12. a, 13. c

Chapter 8

1. T
2. D — wavelength
3. A — frequency
4. B — velocity
5. wavelength
6. T
7. C — pitch
8. A — amplitude
9. A — high
10. F — Both travel at the same speed
11. tension
12. B — lowest
13. A — acoustics
14. A — better
15. B — twice
16. B — decibels
17. T
18. quality
19. A — same speed
20. A — toward

Chapter 9

1. b, 2. c, 3. f, 4. d, 5. g, 6. e
7. h, 8. a, 9. T
10. blue
11. A — reflection
12. B — virtual
13. B — mirror
14. A — convex
15. B — slower
16. A — refraction
17. F — color

Chapter 10

1. T
2. A — electron
3. T
4. charge
5. A — electrostatic
6. B — nonconductor
7. T
8. F — used in golf carts and hybrid cars
9. D — voltage
10. current = voltage/resistance, or in words: current is directly proportional to voltage and inversely proportional to resistance
11. C — resistance
12. voltage

Chapter 11

1. F — Understanding of magnetism was filled with misinformation
2. F — It is drawn toward the north magnetic pole of the earth.
3. F — Magnetic north is 1,100 miles from geographic north.

4. B — gravity
5. T, 6. T
7. C — iron
8. B — repel
9. T, 10. T
11. B — steel
12. A — lost
13. T
14. A — can be turned on and off
15. T
16. D — all of the above

Chapter 12

1. C — James Clerk Maxwell
2. C — the same
3. D — Rudolf Hertz
4. F — FM waves are too short to reflect
5. A — AM
6. C — modulation
7. 3–blue visible light; 1 — AM radio waves; 4 — x-rays; 2 — infrared light
8. D — the second scientific revolution
9. F — color of the light (frequency)
10. T
11. C — photoelectric effect
12. F — mass and velocity
13. T
14. F — Compton effect demonstrates particle nature of light, too.

Chapter 13

1. A — electron
2. B — neutron
3. B — a proton
4. A — fission
5. T
6. C — neutrons
7. D — Lise Meiter
8. T
9. D — slow neutrons
10. A — less
11. F — about 20 percent
12. F — Cold fusion is an unsolved problem.

Chapter 14

1. A — far less
2. F — its frequency
3. A — more
4. T
5. A — an electron
6. T
7. A — fundamental frequency
8. B — an electron
9. T
10. velocity
11. a, 12. d, 13. b, 14. c

Exploring the World of Biology Worksheet Answer Keys

Chapter 1

1. T
2. B. scientists in the 1700s
3. B. chlorophyll
4. C. fungi kingdom
5. F, they can also reproduce by spores
6. T
7. C. release spores
8. A. truffles
9. B. grow and reproduce
10. C. sugar
11. B. mold
12. B. It probably drifted in through an open window.

Chapter 2

1. A. Robert Hooke
2. B. Anton van Leeuwenhoek
3. T

4. A. amoeba
5. F, it can carry out all life functions, but does so in a single cell
6. malaria
7. B. had chlorophyll
8. B. diatoms
9. T
10. fungi
11. false, anaerobic bacteria can survive without oxygen
12. B. nitrogen compounds in the soil
13. f. virus,
    a. animal
    c. fungi
    d. plant
    e. protista
    b. bacteria

Chapter 3

1. Atlantic
2. B. Phoenician
3. C. dogs
4. F, it grew silent
5. F, big dog
6. canine
7. B. mammals
8. A. Guinea along the west coast of Africa
9. T
10. T
11. C. Carl Linneaus
12. B. Species
13. A. disorder or confusion

Chapter 4

1. A. the fall
2. A. as a way to be transported elsewhere
3. almonds
4. A. coconut
5. true
6. vegetative
7. B. do not flower
8. T
9. biennials
10. stem
11. F, sugar (glucose)
12. B. coffee

Chapter 5

1. B. cereal grains.
2. F, in use by Native Americans before then
3. A. oxidation
4. A. fats
5. A. carbohydrate
6. bacteria
7. C. lactose
8. glucose
9. carbon, hydrogen, oxygen
10. T
11. B. heat energy
12. B. fat
13. D. proteins
14. B. proteins

Chapter 6

1. B. mechanical
2. B. grazing animals
3. C. mash food
4. A. sugar
5. T
6. B. chemically changing the food
7. F, milk digests more quickly than a pickle, for instance
8. A. hydrochloric acid
9. B. small intestine
10. C. small intestine
11. A. amino acids
12. A. bitter taste

Chapter 7

1. A. used a compost hotbed

2. B. planting a piece of a potato with an eye in it
3. T
4. California
5. plum
6. A. almond tree
7. Appleseed
8. He was too small and frail for heavy work.
9. B. the Tuskegee Institute in Alabama
10. B. nitrogen
11. cotton
12. A. nitrogen, fixing bacteria grew along their roots
13. F, his dog
14. B. loop

Chapter 8

1. c. insects
   e. spiders and ticks
   b. crabs and lobsters
   a. centipedes
   d. millipedes
2. B. joint
3. A. cricket
4. T
5. F, he observed their behavior in nature
6. head, thorax, abdomen
7. B. mate and reproduce
8. A. butterfly
9. F, Louis Pasteur
10. T
11. A. aphid

Chapter 9

1. B. eight legs and two body segments
2. F, it is relatively harmless
3. A. female black widow
4. B. hunt at night
5. T
6. C. is consistent in thickness without weak spots
7. A. grasshopper
8. to be sure a scorpion is not in them
9. A. to eat at someone else's table
10. A. Lyme disease
11. exoskeleton
12. To grow a larger exoskeleton as they get larger.

Chapter 10

1. F, they continue to be discovered
2. A. invertebrates
3. B. fish
4. D. subphylum
5. A. cold-blooded
6. lateral line
7. B. *The Silent World*
8. C. salamanders
9. skin
10. F, amphibian skin has no scales
11. A. cold-blooded
12. B. decline

Chapter 11

1. T
2. A. cold-blooded
3. A. frog
4. F, reptiles do not have sweat glands
5. A. crocodile
6. to sample airborne chemicals
7. A. heat rays
8. B. keep the victim calm and transport to the hospital
9. B. moccasin
10. D. damages the nervous system
11. B. squeezing their victims
12. A. cold-blooded
13. C. Gila monster
14. chameleon
15. b. copperhead
    f. moccasin
    d. diamondback rattlesnake

c. coral snake
e. king cobra
a. anaconda

Chapter 12

1. B. warm-blooded
2. B. more
3. B. have feathers
4. poultry
5. pigeons
6. C. Roger Tory Peterson
7. D. all of the above
8. B. male
9. They need high-energy food.
10. D. Its feathers have no oil coating.
11. gizzard
12. T
13. g. parrot
    e. falcon
    d. dodo
    h. passenger pigeon
    f. hummingbird
    a. anhinga
    b. Arctic tern
    c. cassowary

Chapter 13

1. A. mammals are cold-blooded
2. T
3. B. seals
4. T
5. they both produce milk for their young
6. bat
7. A. active at night
8. mole
9. T
10. B. rodents
11. B. feline
12. B. chasing them down.
13. A. thick skin
14. T

Chapter 14

1. A. fraud
2. A. past life
3. C. to make the fossil more valuable
4. A. did not match
5. F
6. T
7. C. part of a skull.
8. A. England
9. T
10. D. wise man
11. F; spear points, a flute, and other items have been found.
12. He had to stoop to go in and never thought to look overhead.

Exploring the World of Chemistry — Worksheet Answer Keys

Chapter 1

1. meteorites, 2. D, 3. F, 4. D, 5. A, 6. C
7. carbon, 8. A, 9. B, 10. D, 11. F, 12. B

Chapter 2

1. copper, 2. B, 3. F, 4. F, 5. F, 6. A, 7. A
8. A, 9. mercury, 10. D, 11. T, 12. C
13. mercury, 14. D, 15. A

Chapter 3

1. B, 2. A, 3. T, 4. B, 5. F, 6. sulfuric
7. B, 8. carbon, 9. A, 10. F, 11. T, 12. A
13. dark, 14. F, 15. B, 16. B, 17. B

Chapter 4

1. A, 2. B, 3. D, 4. A, 5. A, 6. T, 7. D
8. D, 9. A, 10. A, 11. B, 12. T, 13. A,
14. B
15. H C N O Cl

16. H2O CO2 HCl

Chapter 5

1. F, 2. B, 3. D, 4. D, 5. C, 6. A, 7. B
8. D, 9. carbon, 10. A, 11. A, 12. B, 13. A
14. D

Chapter 6

1. B, 2. B, 3. D, 4. B, 5. B, 6. D, 7. T
8. A, 9. C, 10. D, 11. F, 12. F, 13. A

Chapter 7

1. C, 2. B, 3. B, 4. T, 5. F, 6. T, 7. A
8. T, 9. A, 10. F, 11. F, 12. A, 13. T
14. C, 15. B

Chapter 8

1. B, 2. B, 3. T, 4. F, 5. F, 6. C, 7. atoms
8. A, 9. B, 10. A, 11. C, 12. A, 13. B
14. B

Chapter 9

1. B, 2. A, 3. D, 4. C, 5. Sahara, 6. F
7. C, 8. B, 9. B, 10. B, 11. Tyre, 12. B

Chapter 10

1. B, 2. T, 3. C, 4. A, 5. A, 6. B, 7. B
8. B, 9. F, 10. B, 11. B, 12. B, 13. A

Chapter 11

1. four, 2. A, 3. C, 4. B, 5. T, 6. B, 7. four
8. F, 9. T, 10. T, 11. F, 12. T, 13. B, 14. D
15. B, 16. A, 17. B, 18. C

Chapter 12

1. B, 2. A, 3. carbon, 4. F, 5. F, 6. A, 7. B
8. D, 9. C, 10. F, 11. F, 12. D, 13. A
14. F, 15. A

Chapter 13

1. nitrogen, 2. B, 3. T, 4. F, 5. nitrogen
6. D, 7. A, 8. B, 9. B, 10. T, 11. C, 12. F
13. D, 14. A, 15. T, 16. F, 17. T

Chapter 14

1. B, 2. A, 3. B, 4. T, 5. B, 6. D, 7. A
8. T, 9. F, 10. B, 11. T, 12. C, 13. B

Chapter 15

1. F, 2. T, 3. A, 4. A, 5. F, 6. A, 7. F
8. T, 9. T, 10. A, 11. C, 12. T, 13. F
14. A, 15. T

Chapter 16

1. b, 2. c, 3. g, 4. d, 5. e, 6. a, 7. f, 8. i
9. h, 10. j

Unit One Quiz, chapters 1–4

1. c. earth rotates on its axis once
2. e. seven days
3. d. moon revolves around the earth once
4. a. due to the tilt of the earth's axis, equal to three months
5. b. earth revolves around the sun once
6. 5,280
7. pound
8. A. authorized by Julius Caesar
9. D. Romans
10. C. 2400
11. D. trains,
12. B. engineers used two different measures of force
13. C. drugs
14. B. human body,
15. C. the distance between two scratch marks
16. D. 212 degrees
17. (divide 360 by 3, 5, 20, and 60), a. 120 days, b. 72 days, c. 18 days, d. 6 days
18. a. 60 inches (multiplying 15 hands by 4 inches per hand gives 60 inches)
    b. 5 feet (60 inches is equal to 5 feet: 60 in ÷ 12 in. per ft. = 5 ft.)
19. So the calendar will match the seasons; or so the calendar year will be the same length as the solar year.
20. 6:30 a.m. The second watch began at 4:00 a.m. Each bell is ½ hour. Five bells are 2½ hours: 4:00 a.m. + 2 hr. 30 min. = 6:30 a.m.
21. one hour later, 4:00 p.m. MST is 5:00 p.m. CST
22. 5.499 miles or about 5.5 miles. Divide 29,035 feet by the conversion factor of 5,280 feet per mile.
23. Answer varies depending on actual heart rate. For 72 beats per min: 72 beats per min. x 60 min. per hr. x 24 hr. per day = 103,680 beats per day
24. Answer varies. Multiply weight in pounds by the conversion factor of 16 ounces per pound.

25. B, 26. A, 27. A, 28. B, 29. A

Unit Two Quiz, chapters 5–8

1. e. is not a polygon
2. a. a polygon with five sides
3. d. a quadrilateral with opposite sides parallel and equal in length
4. c. a polygon with three sides and one right angle
5 b. a rectangle with four equal sides
6. d. perimeter of a rectangle
7. e. perimeter of a square
8. f. volume
9. b. area of a rectangle
10. a. area of a circle
11. c. area of a square
12. e. ancient Greek who worked out a way to show large numbers that he called myriads
13. d. wrote Elements of Geometry
14. c. proved planets follow elliptical orbits
15. a. discovered that the sum of the 3 angles of any triangle is 180 degrees
16. b. used ratios to find the heights of buildings
17. b. all points are the same distance from the center
18. d. the orbit of Halley's comet is of this shape
19. a. a mirror of this shape will focus sunlight
20. c. the first part of the name means over or beyond
21. b. palindromes
22. c. prime numbers
23. a. Fibonacci numbers 8
24. d. square numbers
25. e. triangular numbers
26. squares, square
27. 233 = 89 + 144
28. D. a triangle with a right angle
29. D. eight times as great
30. D. an ellipse
31. C. number theory

32. F

33. F

34. F

35. F

36. F

37. 140 tiles. The area of the room is 140 square feet, A = L x W = 14 ft. x 10 ft. = 140 sq. ft., and each tile covers one square foot, so 140 tiles are needed.

38. 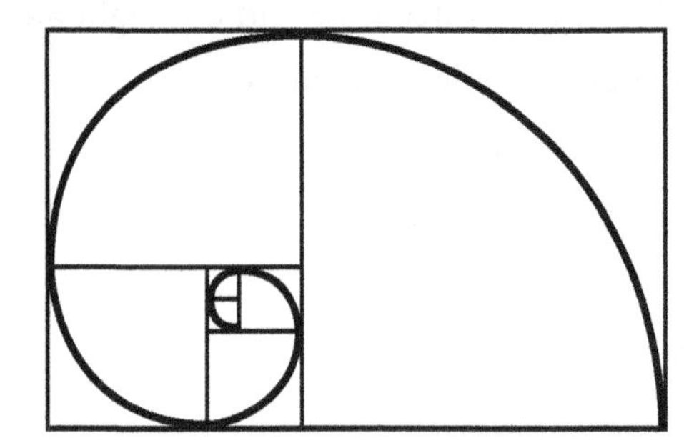

Unit Three Quiz, chapters 9–11

1. a. counting numbers

2. g. whole numbers

3. b. even numbers

4. e. odd numbers

5. c. integers

6. f. rational numbers

7. d. irrational numbers

8. b, 9. d, 10. c, 11. a, 12. b

13. d, 14. a, 15. c

16. e. proved that xn + yn = zn has no solution with whole numbers except for n = 2

17. b. he called his triangle an arithmetic triangle

18. a. discovered how to calculate the coefficients of a binomial raised to a power

19. d. solved the Konigsberg bridge problem

20. c. his last theorem was solved in 1995

21. D. division

22. C. 8

23. B. 25

24. A. denominator

25. B. an irrational number

26. D. all of the above

27. C. solving

28. A. a coefficient

29. B. geometry

30. C. four-color map problem

31. B. factorial

32. C. 120

33. 17,576,000 — Any one of 26 letters (A through Z) can be chosen to fill the first three positions. Any one of 10 digits (zero through nine) can be chosen to fill the second group of three positions: 26 x 26 x 26 x 10 x 10 x 10 = 17,576,000.

34. a. Utah, b. Colorado, c. Arizona, d. New Mexico

Unit Four Quiz, chapters 12–14

1. h. wrote the first computer program

2. c. built first general purpose calculating machine

3. d. built the difference engine

4. f. invented tabulating machines used in the 1890 census

5. g. spent six years calculating the orbit of Mars,

6. a. built a calculator called the Step Reckoner

7. e. invented logarithms

8. b. built a calculator to help his father, a tax collector

9. three, ten, 0.477

10. 9

11. B. 1

12. D. 5.28 x 103

13. B. punched cards

14. A. random access memory

15. A. bit

16. C. eight

17. D. the three colors of red, green, and blue

18. D. transmitting only those pixels that are different from the previous one

19. B. months

20. A. electric signals can go no faster than the speed of light

21. central processing unit

22. About 3.8 seconds: 26,747 characters / 7,000 bytes (characters) per second = 3.821 seconds

23. Samson's Riddle: Judges 14:14 tells how Samson came to think of the puzzle. He found a lion's carcass with a honeycomb from a beehive inside. Out of the eater (lion), something to eat, out of the strong, something sweet (honey).
24. On the road to St. Ives: Only one person is on the road to St. Ives. Instead of a large number, this is a trick question. The last line of the rhyme asks how many were going to St. Ives. The wives and the things they carried were going away from St. Ives. The narrator (the "I" person) is the one going to St. Ives.
25. River Crossing: First, take the goat across and leave him on the far bank. The carrots will be safe left alone with the wolf. Return for the carrots and leave them on the far bank but pick up the goat and return with him to the near bank. Leave the goat and paddle the wolf across. Leave the wolf with the carrots and return for the goat.

Exploring the World of Physics — Quiz Answer Keys

Unit One Quiz, chapters 1–4

1. e. If f = 0 then a = 0
2. a. a = f/m
3. c. fab = -fba
4. b. f = m x a
5. d. I = f x t
6. f. p = m x v
7. time
8. 53 miles per day. Average speed is the distance divided by the time: speed = distance/time = 3,710 miles/70 days = 53 mi./da.
9. 31.8 ft./sec. v1 = a x t = (5.3 ft./sec^2) (6 sec.) = 31.8 ft./sec.
10. Velocity
11. product, square
12. M.A. = 9
13. C. matter
14. C. a vacuum
15. D. rolled them down a ramp
16. D. change in speed by the change in time
17. D. velocity
18. B. elliptical
19. A. Archimedes
20. B. a lever
21. 7.5 mi./hr. x sec. Acceleration = (change in speed)/(change in time) = (60 mi./hr.) (8 sec.) = 7.5 mi./hr. x sec.
22. The acceleration of an object is directly proportional to the force acting on it and inversely proportional to its mass.
23. To every action there is an equal and opposite reaction.
24. A. forward thrust of propeller
    B. drag
    C. upward lift of wings
    E. downward pull of gravity
25. A. Fulcrum
    B. Load
    C. Effort Points
26. A. Fulcrum
    B. Load
    C. Effort Points

Unit Two Quiz, chapters 5–7

1. b. The lifting force acting on a solid object immersed in water is equal to the weight of the water shoved aside by the object.
2. d. The volume of a gas is inversely proportional to the pressure.
3. a. Pressure times volume of any gas divided by the temperature is a constant.
4. c. The velocity of a fluid and its pressure are inversely related.
5. energy
6. heat
7. temperature
8. 0.058 or about six percent.
   Efficiency = (T1 – T2)/T1 = (291 K – 274 K)/

(291 K) = 17/291 = 0.058 or less than six percent

9. D. work
10. D. work
11. A. energy
12. C. mercury
13. C. Daniel Fahrenheit
14. C. volume
15. C. horsepower and watt
16. Almost every time that energy changes form, the amount of A. heat B. kinetic C. potential energy increases.
17. A. conduction
18. B. convection
19. B. square root
20. F — only since the 1800s
21. T
22. F — The desk must move for work to be done.
23. T
24. F — It becomes four times as great.
25. F — They can measure higher temperatures with electrical conductivity and color of light emitted by glowing substance.
26. T, 27. T, 28. T, 29. T
30. F — Because it will snap back to its original shape.
31. T, 32. T
33. You can measure heat capacity by heating identical masses of different substances and then placing them on a block of ice. Observe the different distances they melt into the ice.

Unit Three Quiz, chapters 8–10

1. b. Cornea
2. c. Iris
3. f. Pupil
4. d. Lens
5. g. Retina
6. e. Optic nerve
7. h. Rods
8. a. Cones
9. tension
10. quality
11. blue
12. charge
13. voltage
14. B. lowest
15. A. better
16. A. same speed
17. A. toward
18. A. reflection
19. B. virtual
20. B. mirror
21. A. convex
22. B. slower
23. refraction
24. A. electron
25. A. electrostatic
26. B. nonconductor
27. D. wavelength
28. C. pitch
29. A. acoustics
30. B. decibels
31. D. voltage
32. C. resistance
33. current = voltage/resistance, or in words: current is directly proportional to voltage and inversely proportional to resistance.
34. T, 35. T
36. F — color
37. T, 38. T, 39. T
40. F — used in golf carts and hybrid cars
41. rustling of leaves
42. whisper
43. average home sounds (such as the humming of a refrigerator)
44. automobile
45. ordinary conversation
46. heavy street traffic
47. jack hammer
48. thunder

Unit Four Quiz, chapters 11–14

1. a. Developed a model of the atom and electron orbits
2. d. Proposed matter waves and calculated their wavelengths
3. b. Explained black body radiation by using energy quanta
4. c. Developed the uncertainty principle
5. velocity
6. A. can be turned on and off
7. D. all of the above
8. C. James Clerk Maxwell
9. D. Rudolf Hertz
10. C. modulation
11. D. the second scientific revolution
12. C. photoelectric effect
13. A. an electron
14. C. neutrons
15. D. Lise Meiter
16. D. slow neutrons
17. B. an electron
18. B. steel
19. A. lost
20. A. AM
21. A. fission
22. A. less
23. A. far less
24. A. more
25. A. an electron
26. A. fundamental frequency
27. 3 — blue visible light; 1 — AM radio waves; 4 — x-rays; 2 — infrared light
28. A. nucleus
    B. electron
    C. proton
    D. neutron
    E. quark

Exploring the World of Biology Quiz Answer Keys

Unit One Quiz, chapters 1–3

1. f. virus
   a. animal
   c. fungi .
   d. plant
   e. protista
   b. bacteria
2. malaria
3. fungi
4. Atlantic
5. canine
6. B. chlorophyll
7. C. fungi kingdom
8. C. release spores
9. C. sugar
10. B. mold
11. A. amoeba
12. B. diatoms
13. B. Phoenician
14. C. dogs
15. A. Guinea along the west coast of Africa
16. C. Carl Linnaeus
17. B. grow and reproduce
18. B. Anton van Leeuwenhoek
19. B. had chlorophyll
20. B. mammals.
21. kingdom: animal
22. phylum: chordata
23. class: mammalia
24. order: carnivore
25. family: feline
26. genus: puma
27. species: concolor

Unit Two Quiz, chapters 4–7

1. almonds
2. vegetative
3. biennials
4. stem
5. bacteria

6. glucose
7. carbon, hydrogen, oxygen
8. California
9. plum
10. cotton
11. A. as a way to be transported elsewhere
12. A. coconut
13. A. carbohydrate
14. C. lactose
15. D. proteins
16. C. mash food
17. B. chemically changing the food
18. C. small intestine
19. A. amino acids
20. A. bitter taste
21. B. nitrogen
22. A. nitrogen, fixing bacteria grew along their roots
23. A. the fall B. early spring
24. do not flower
25. B. small intestine
26. A. used a compost hotbed
27. B. planting a piece of a potato with an eye in it
28. A. almond tree
29. He was too small and frail for heavy work.
30. esophagus
31. cardia
32. longitudinal muscle layer
33. circular muscle layer
34. pyloric sphincter
35. duodenum
36. body
37. oblique muscle layer
38. rugae

Unit Three Quiz, chapters 8–11

1. c. insects
   e. spiders and ticks
   b. crabs and lobsters
   a. centipedes
2. b. copperhead
   f. moccasin
   d. diamondback rattlesnake
   c. coral snake
   e. king cobra
   a. anaconda
3. exoskeleton
4. skin
5. chameleon
6. A. female black widow
7. C. is consistent in thickness without weak spots
8. A. grasshopper
9. B. fish
10. D. subphylum
11. C. salamanders
12. B. joint
13. A. cricket
14. B. mate and reproduce
15. B. eight legs and two body segments
16. B. hunt at night
17. A. to eat at someone else's table
18. A. cold-blooded
19. A. cold-blooded
20. B. decline
21. To grow a larger exoskeleton as they get larger.
22. lateral line
23. to sample airborne chemicals
24. primary venom duct
25. venom gland
26. compressor muscle
27. secondary venom duct
28. fang
29. accessory gland

Unit Four Quiz, chapters 12–14

1. g. parrot
   e. falcon
   d. dodo
   h. passenger pigeon
   f. hummingbird
   a. anhinga

b. Arctic tern

c. cassowary

2. poultry
3. gizzard
4. bat
5. mole
6. A. mammals are cold-blooded
7. C. to make the fossil more valuable
8. C. part of a skull
9. A. England
10. D. wise man
11. B. warm-blooded
12. B. more
13. B. seals
14. A. active at night
15. B. rodents
16. B. feline
17. B. chasing them down
18. A. fraud
19. A. past life
20. A. did not match
21. T, 22. T, 23. F, 24. T, 25. T
26. F — spear points, a flute, and other items have been found.
27. pigeons
28. They need high-energy food.
29. They both produce milk for their young.
30. He had to stoop to go in and never thought to look overhead.

Exploring the World of Chemistry Quiz Answer Keys

Unit One Quiz, chapters 1–4

1. meteorites
2. carbon
3. copper
4. mercury
5. sulfuric
6. carbon
7. Dark
8. H2O, CO2, HCl
9. D. wood that has been heated without oxygen
10. D. wrought iron
11. B. heating it in an oven for several days
12. B. gold
13. A. copper
14. A. copper alloy
15. A. copper
16. D. quicksilver
17. B. heated sulfur with raw rubber
18. B. communicate new ideas rapidly
19. B. hydrochloric acid
20. D. water generator
21. A. a violent explosion results
22. D. oxygen
23. F, 24. F, 25. T, 26. T
27. H hydrogen
28. C carbon
29. N nitrogen
30. O oxygen
31. Cl chlorine

Unit Two Quiz, chapters 5–8

1. carbon
2. atoms
3. C. battery to produce electric current
4. A. electrons
5. B. Humphry Davy
6. D. electricity from a strong battery
7. B. Russia
8. D. to train teachers
9. B. help his students
10. D. writing information about the elements on note cards
11. D. predicted the properties of three missing elements
12. C. spectroscope
13. A. cesium
14. A. an electric spark

15. C. the sun
16. B. inert gases
17. A. because he enjoyed it
18. B. measures the strength of radiant energy
19. C. passed straight through
20. A. bounced out in the same direction it entered
21. F, 22. T, 23. F, 24. T, 25. T, 26. T
27. T, 28. F

Unit Three Quiz, chapters 9–12

1. Sahara
2. Tyre
3. four
4. four
5. Carbon
6. B. stop an outbreak of typhoid fever
7. B. electrical attraction
8. B. sharing
9. B. contracts
10. A. floats
11. B. poorly
12. B. slowly
13. B. water
14. B. water
15. B. long
16. B. slick
17. B. isomers
18. B. found in the nonliving environment
19. A. formed by living organisms
20. B. has more protons than electrons
21. A. attract one another
22. D. salt makers
23. C. an oxygen atom face with hydrogen atom ears
24. B. it is fed by melting snow in the mountains
25. B. 100°C [212°F]
26. A. marsh
27. C. hydrogen
28. D. shellac
29. A. a substance that clogged test tubes
30. F, 31. T, 32. F, 33. T, 34. T, 35. F
36. T, 37. F, 38. F, 39. F, 40. F

Unit Four Quiz, chapters 13–16

1. b. Robert Boyle
2. c. Henry Cavendish
3. g. Antoine Laurent Lavoisier
4. d. John Dalton
5. e. Humphry Davy
6. a. Jöns Jakob Berzelius
7. f. Michael Faraday
8. i. Louis Pasteur
9. h. Dmitri Ivanovich Mendeleev
10. j. William Ramsay
11. nitrogen
12. nitrogen
13. D. strong cell walls
14. C. a railroad
15. D. silica
16. A. for safe detonation of explosives
17. A. diamond
18. B. to hide the fact that glass makers could not make clear glass
19. D. all of the above
20. A. a diamond
21. C. semiconductor
22. A. aluminum
23. A. aluminum
24. A. aluminum
25. C. pitchblende
26. A. americium
27. B. gunpowder
28. B. unpredictable and capable of exploding without warning
29. B. keep it secret
30. B. second most abundant
31. quartz
32. B. silicone oil
33. B. silicon
34. A. planet Uranus

1. c. earth rotates on its axis once
2. e. seven days
3. d. moon revolves around the earth once
4. a. due to the tilt of the earth's axis, equal to three months
5. b. earth revolves around the sun once
6. b. all points are the same distance from the center
7. d. the orbit of Halley's comet is of this shape
8. a. a mirror of this shape will focus sunlight
9. c. the first part of the name means over or beyond
10. a. counting numbers
11. g. whole numbers
12. b. even numbers
13. e. odd numbers
14. c. integers
15. f. rational numbers
16. d. irrational numbers
17. h. wrote the first computer program
18. c. built first general purpose calculating machine
19. d. built the difference engine
20. f. invented tabulating machines used in the 1890 census
21. g. spent six years calculating the orbit of Mars
22. a. built a calculator called the Step Reckoner
23. e. invented logarithms
24. b. built a calculator to help his father, a tax collector
25. 5,280
26. pound
27. squares, square
28. 233 = 89 + 144
29. three, ten, 0.477
30. 9
31. B. engineers used two different measures of force
32. C. drugs
33. D. the distance light travels in 1/299,792,458 of a second
34. D. 212 degrees
35. D. a triangle with a right angle
36. D. eight times as great
37. D. an ellipse
38. C. number theory
39. A. denominator
40. B. an irrational number
41. B. geometry
42. four-color map problem
43. D. the three colors of red, green, and blue
44. D. transmitting only those pixels that are different from the previous one
45. B. months
46. A. electric signals can go no faster than the speed of light
47. a. 60 inches (multiplying 15 hands by 4 inches per hand gives 60 inches)
    b. 5 feet (60 inches is equal to 5 feet: 60 in. ÷ 12 in. per ft. = 5 ft.)
48. So the calendar will match the seasons; or so the calendar year will be the same length as the solar year.
49. one hour later, 4:00 p.m. MST is 5:00 p.m. CST
50. 5.499 miles or about 5.5 miles. Divide 29,035 feet by the conversion factor of 5,280 feet per mile.
51. 140 tiles. The area of the room is 140 square feet, A = L x W = 14 ft. x 10 ft. = 140 sq. ft., and each tile covers one square foot, so 140 tiles are needed.
52. Sock Puzzle: One extra sock is enough. If the socks he is wearing match, he does not need the spare one. If his socks are not alike, then one will be the same color as the one he is carrying, giving him a matching pair.
53. Durer's Number Square: To get started, you should figure out the sum for each row, column, and diagonal. The numbers 1 through 9 sum to 45, and each row (and column and diagonal) must sum to 15 (45/3 = 15). One possible arrangement is:

8	1	6
3	5	7
4	9	2

The other seven solutions are merely rotations and mirror images of this solution.

Exploring the World of Physics — Test Answer Key

1. e. If f = 0 then a = 0
2. a. a = f/m
3. c. fab = -fba
4. b. f = m x a
5. d. I = f x t
6. f. p = m x v
7. b. The lifting force acting on a solid object immersed in water is equal to the weight of the water shoved aside by the object.
8. d. The volume of a gas is inversely proportional to the pressure.
9. a. Pressure times volume of any gas divided by the temperature is a constant.
10. c. The velocity of a fluid and its pressure are inversely related.
11. b. Cornea
12. c. Iris
13. f. Pupil
14. d. Lens
15. g. Retina
16. e. Optic nerve
17. h. Rods
18. a. Cones
19. a. Developed a model of the atom and electron orbits
20. d. Proposed matter waves and calculated their wavelengths
21. b. Explained black body radiation by using energy quanta
22. c. Developed the uncertainty principle
23. time
24. Velocity
25. heat
26. temperature
27. quality
28. voltage
29. velocity
30. C. matter
31. C. a vacuum
32. D. change in speed by the change in time
33. D. velocity
34. A. energy
35. C. mercury
36. C. Daniel Fahrenheit
37. C. volume
38. C. pitch
39. A. acoustics
40. B. decibels
41. D. voltage
42. C. James Clerk Maxwell
43. D. Rudolf Hertz
44. D. Lise Meiter
45. D. slow neutrons
46. The acceleration of an object is directly proportional to the force acting on it and inversely proportional to its mass.
47. To every action there is an equal and opposite reaction.
48. current = voltage/resistance, or in words: current is directly proportional to voltage and inversely proportional to resistance.
49. T
50. F — It becomes four times as great
51. F — They can measure higher temperatures with electrical conductivity and color of light emitted by glowing substance.

52. T , 53. T , 54. T

55. A. forward thrust of propeller
    B. drag
    C. upward lift of wings
    E. downward pull of gravity
56. 3 — blue visible light; 1 — AM radio waves; 4 — x-rays; 2 — infrared light

1. f. virus
   a. animal
   c. fungi
   d. plant
   e. protista
   b. bacteria
2. c. insects
   e. spiders and ticks
   b. crabs and lobsters
   a. centipedes
3. b. copperhead
   f. moccasin
   d. diamondback rattlesnake
   c. coral snake
   e. king cobra
   a. anaconda
4. g. parrot
   e. falcon
   d. dodo
   h. passenger pigeon
   f. hummingbird
   a. anhinga
   b. Arctic tern
   c. cassowary
5. malaria
6. fungi
7. Atlantic
8. canine
9. almonds
10. vegetative
11. biennials
12. stem
13. bacteria
14. glucose
15. carbon, hydrogen, oxygen
16. California
17. exoskeleton
18. gizzard
19. bat
20. C. release spores
21. C. sugar
22. B. mold
23. A. amoeba
24. A. amino acids
25. A. bitter taste
26. B. nitrogen
27. A. nitrogen, fixing bacteria grew along their roots
28. To grow a larger exoskeleton as they get larger.
29. to sample airborne chemicals
30. pigeons
31. They need high-energy food.
32. they both produce milk for their young
33. He had to stoop to go in and never thought to look overhead.
34. kingdom: animal
35. phylum: chordata
36. class: mammalia
37. order: carnivore
38. family: feline
39. genus: puma
40. species: concolor

Exploring the World of Chemistry Test Answer Key

1. b. Robert Boyle
2. c. Henry Cavendish
3. g. Antoine Laurent Lavoisier
4. d. John Dalton.
5. e. Humphry Davy
6. a. Jöns Jakob Berzelius
7. f. Michael Faraday
8. i. Louis Pasteur
9. h. Dmitri Ivanovich Mendeleev
10. j. William Ramsay
11. meteorites
12. carbon
13. copper
14. mercury
15. sulfuric
16. carbon
17. Dark
18. H2O, CO2, HCl
19. carbon
20. atoms
21. Sahara
22. Tyre
23. four
24. four
25. Carbon
26. nitrogen
27. nitrogen
28. A. copper
29. D. quicksilver
30. B. heated sulfur with raw rubber
31. B. communicate new ideas rapidly
32. D. predicted the properties of three missing elements
33. C. spectroscope
34. A. cesium
35. A. an electric spark
36. A. attract one another
37. D. salt makers
38. C. an oxygen atom face with hydrogen atom ears
39. B. it is fed by melting snow in the mountains
40. D. all of the above
41. A. a diamond
42. C. semiconductor
43. A. aluminum
44. B. stop an outbreak of typhoid fever
45. B. electrical attraction
46. B. sharing
47. B. second most abundant
48. B. silicon

49. F, 50. F, 51. T, 52. T, 53. F, 54. T
55. F, 56. T, 57. T, 58. T, 59. T, 60. F
61. F, 62. T, 63. F, 64. T, 65. T, 66. F
67. T, 68. F, 69. F, 70. F, 71. F

72. H hydrogen, C carbon, N nitrogen, O oxygen, Cl chlorine (see Periodic Table on following page)

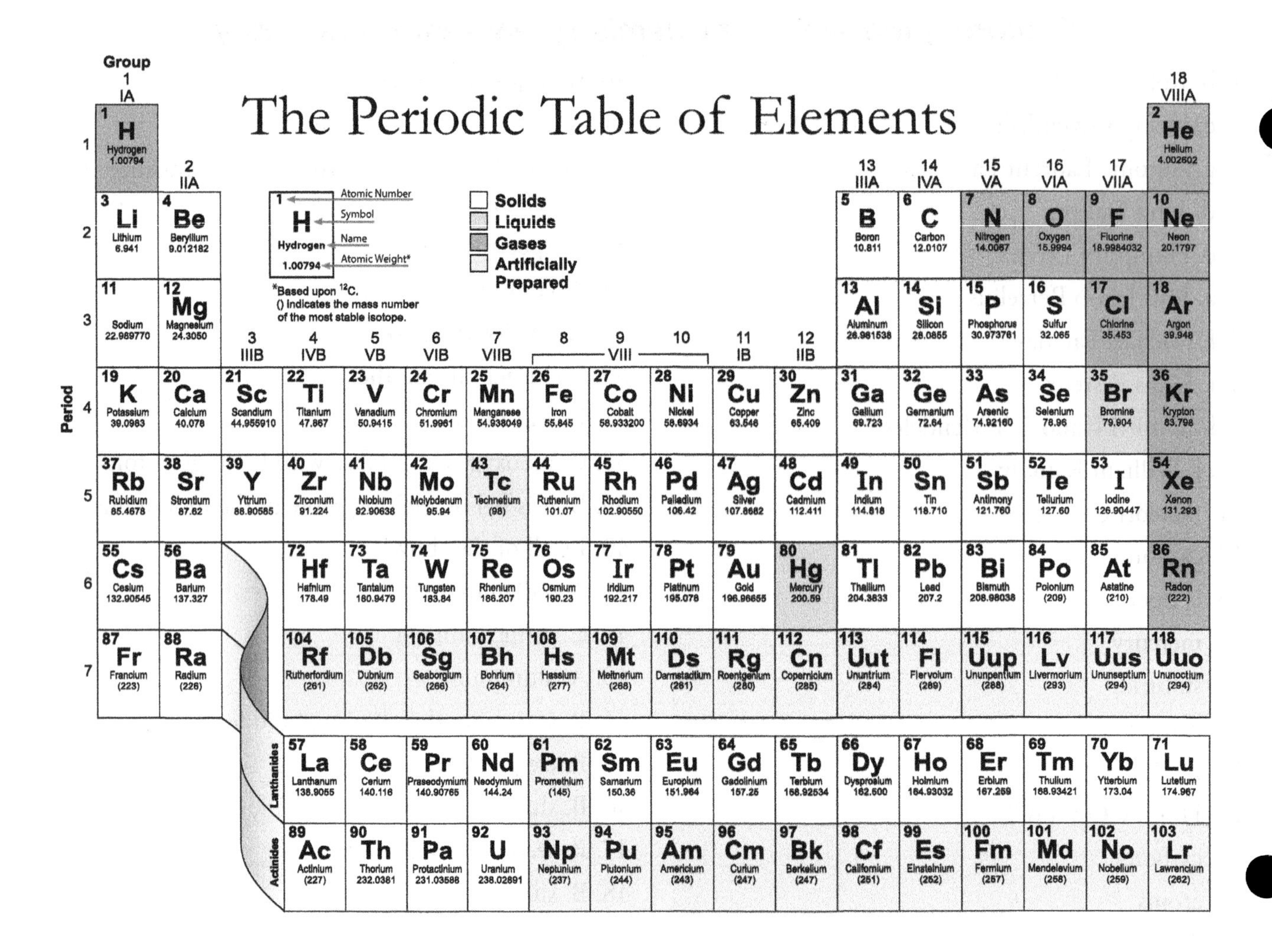

The Periodic Table of Elements
Group
1 IA
2 IIA
3 IIIB
4 IVB
5 VB
6 VIB
7 VIIB
8 9 10 VIII
11 IB
12 IIB
13 IIIA
14 IVA
15 VA
16 VIA
17 VIIA
18 VIIIA
Period
1 2 3 4 5 6 7
Atomic Number
Symbol
Name
Atomic Weight*
1 H Hydrogen 1.00794
*Based upon ¹²C.
() indicates the mass number of the most stable isotope.
Solids
Liquids
Gases
Artificially Prepared
1 H Hydrogen 1.00794
2 He Helium 4.002602
3 Li Lithium 6.941
4 Be Beryllium 9.012182
5 B Boron 10.811
6 C Carbon 12.0107
7 N Nitrogen 14.0067
8 O Oxygen 15.9994
9 F Fluorine 18.9984032
10 Ne Neon 20.1797
11 Na Sodium 22.989770
12 Mg Magnesium 24.3050
13 Al Aluminum 26.981538
14 Si Silicon 28.0855
15 P Phosphorus 30.973761
16 S Sulfur 32.065
17 Cl Chlorine 35.453
18 Ar Argon 39.948
19 K Potassium 39.0983
20 Ca Calcium 40.078
21 Sc Scandium 44.955910
22 Ti Titanium 47.867
23 V Vanadium 50.9415
24 Cr Chromium 51.9961
25 Mn Manganese 54.938049
26 Fe Iron 55.845
27 Co Cobalt 58.933200
28 Ni Nickel 58.6934
29 Cu Copper 63.546
30 Zn Zinc 65.409
31 Ga Gallium 69.723
32 Ge Germanium 72.64
33 As Arsenic 74.92160
34 Se Selenium 78.96
35 Br Bromine 79.904
36 Kr Krypton 83.798
37 Rb Rubidium 85.4678
38 Sr Strontium 87.62
39 Y Yttrium 88.90585
40 Zr Zirconium 91.224
41 Nb Niobium 92.90638
42 Mo Molybdenum 95.94
43 Tc Technetium (98)
44 Ru Ruthenium 101.07
45 Rh Rhodium 102.90550
46 Pd Palladium 106.42
47 Ag Silver 107.8682
48 Cd Cadmium 112.411
49 In Indium 114.818
50 Sn Tin 118.710
51 Sb Antimony 121.760
52 Te Tellurium 127.60
53 I Iodine 126.90447
54 Xe Xenon 131.293
55 Cs Cesium 132.90545
56 Ba Barium 137.327
72 Hf Hafnium 178.49
73 Ta Tantalum 180.9479
74 W Tungsten 183.84
75 Re Rhenium 186.207
76 Os Osmium 190.23
77 Ir Iridium 192.217
78 Pt Platinum 195.078
79 Au Gold 196.96655
80 Hg Mercury 200.59
81 Tl Thallium 204.3833
82 Pb Lead 207.2
83 Bi Bismuth 208.98038
84 Po Polonium (209)
85 At Astatine (210)
86 Rn Radon (222)
87 Fr Francium (223)
88 Ra Radium (226)
104 Rf Rutherfordium (261)
105 Db Dubnium (262)
106 Sg Seaborgium (266)
107 Bh Bohrium (264)
108 Hs Hassium (277)
109 Mt Meitnerium (268)
110 Ds Darmstadtium (281)
111 Rg Roentgenium (280)
112 Cn Copernicium (285)
113 Uut Ununtrium (284)
114 Fl Flerovium (289)
115 Uup Ununpentium (288)
116 Lv Livermorium (293)
117 Uus Ununseptium (294)
118 Uuo Ununoctium (294)
Lanthanides
57 La Lanthanum 138.9055
58 Ce Cerium 140.116
59 Pr Praseodymium 140.90765
60 Nd Neodymium 144.24
61 Pm Promethium (145)
62 Sm Samarium 150.36
63 Eu Europium 151.964
64 Gd Gadolinium 157.25
65 Tb Terbium 158.92534
66 Dy Dysprosium 162.500
67 Ho Holmium 164.93032
68 Er Erbium 167.259
69 Tm Thulium 168.93421
70 Yb Ytterbium 173.04
71 Lu Lutetium 174.967
Actinides
89 Ac Actinium (227)
90 Th Thorium 232.0381
91 Pa Protactinium 231.03588
92 U Uranium 238.02891
93 Np Neptunium (237)
94 Pu Plutonium (244)
95 Am Americium (243)
96 Cm Curium (247)
97 Bk Berkelium (247)
98 Cf Californium (251)
99 Es Einsteinium (252)
100 Fm Fermium (257)
101 Md Mendelevium (258)
102 No Nobelium (259)
103 Lr Lawrencium (262)

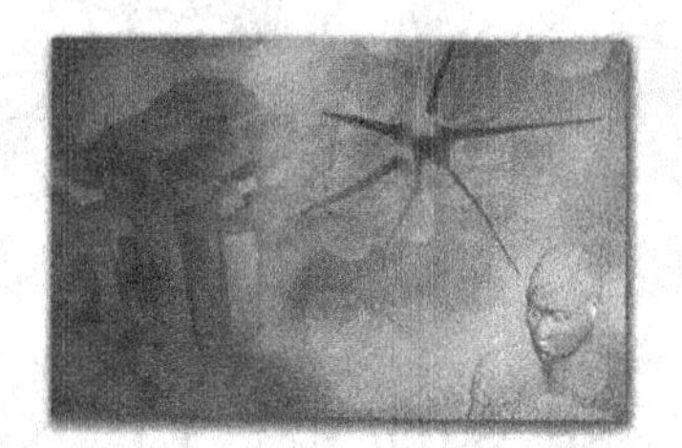

8-1/2 x 11 Paperback • 160 pages • b&w interior
ISBN-13: 978-0-89051-552-5

The Top-selling "Exploring" Series

A great evolution-free resource tool!

This series provides a solid foundation of fact for each subject. Concise information, along with chapter questions, illustrations, and photos to emphasize the facts builds a strong foundation for understanding their respective topic from a Christian world view. Each book includes over 100 illustrations, charts, and photos along with key facts, terms, definitions, chapter review questions, and answer key.

- ***Exploring the World of Biology*** From Mushrooms to Complex Life Forms.
- ***Exploring Planet Earth*** uncovers the history of civilization, historical people, and places.
- ***Exploring the History of Medicine*** examines modern medicine from ancient Greeks to today.
- ***Exploring the World Around You*** tours the planet and its seven biomes.
- ***Exploring the World of Mathematics*** traces the history of mathematic principles and theories.
- ***Exploring the World of Physics*** captures the workings of simple machines to nuclear energy.
- ***Exploring the World of Chemistry*** investigates ancient metals to high-speed computers.

A topic-specific, information-rich, evolution-free series from Master Books

The World of Chemistry
ISBN-13: 978-0-89051-295-1

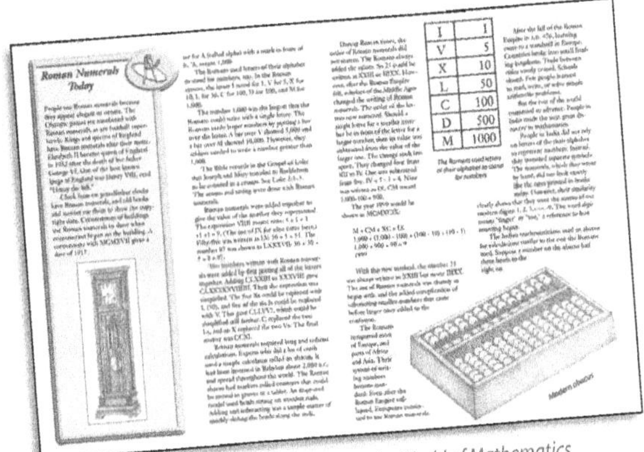

sample interior from Exploring the World of Mathematics

The World of Physics
ISBN-13: 978-0-89051-466-5

The World Around You
ISBN-13: 978-0-89051-377-4

The World of Mathematics
ISBN-13: 978-0-89051-412-2

Planet Earth
ISBN-13: 978-0-89051-178-7

The History of Medicine
ISBN-13: 978-0-89051-248-7